Technology of Terry Weaving

Selamu Temesgen,

Head, Department of Textile Engineering,

Kombolcha Institute of Technology,

Wollo University,

Kombolcha, Ethiopia.

Dr.M. Bharani,

Assistant Professor,

Department of Textile Engineering,

Kombolcha Institute of Technology,

Wollo University,

Kombolcha, Ethiopia.

Published by

Technology of Terry Weaving

Copyright © 2017 by Bonfring

ISBN 978-93-86638-18-2

Authors

Selamu Temesgen

Dr.M. Bharani

Bonfring

309, 2nd Floor, 5th Street Extension, Gandhipuram,

Coimbatore-641 012.

Tamilnadu, India.

E-mail: info@bonfring.org

Website: www.bonfring.org

Phone: 0422 4213231

Preface

In the present information and knowledge era, knowledge has become a key resource. The Technology of Terry weaving has undergone a considerable innovation, aiming to improve productivity and quality. With similar objectives, the new methods of Terry fabric production were explored and developed.

Amongst these, the most prominent ones are Terry towels with various quality variation. The demands on quality of Terry fabrics are increasing along with the rise in volumes of their requirement. In this context, this book is an attempt to create an awareness on recent innovations in Terry fabric manufacturing and summarise the present status and future prospects of Terry fabric manufacturing industry by providing latest and organised information.

Further, This book deals about Quality aspects, dyeing procedure as per industry practice. In addition to that, the book deals about the various defects and remedial procedures terry fabric. It includes various articles as per the research of Academicians, Industry Experts, Researchers and Intellectual Students from various parts of the world, given by prominent machine manufacturers and distinguished industry personnel.

We are immensely pleased to record our sincere thanks to our colleagues, staff members and student volunteers for their assistance in preparing this book.

Selamu Temesgen
Dr.M. Bharani

Chapter	**Contents**	**Page No**

I	Introduction	1
	1. Introduction	1
	2. History of Terry Weaving	2
II	Application of Natural Fibres in Terry Towel Manufacturing	3
	1. Introduction	3
	2. Cotton Fibers	3
	3. Bamboo Fiber	4
	4. Hemp Fiber	5
	5. Wood Fiber	6
III	Parts and Classification of Terry Towels	7
	1. Parts of a Conventional Terry	7
	2. Classification of Terry Towels	7
IV	Structure of Terry Towel and Terry Fabric Weaving	9
	1. Fibers Used in Terry Towels	9
	2. Other Fibers	9
	3. Yarns which are Used in Terry Towels	9
	4. Construction of Terry Towels	11
	5. Loop Formation Technology	13
	6. Steps of Terry Weaving	13
	7. Basic Movements	14
	8. The Shedding Mechanism	14
	9. Filling Insertion	15
	10. Beat-up	16
	11. Complementary Motions	18
	12. Auxiliary Motions	21
V	Technology of Terry Towel Production	23
	1. Terry Weaving	23
	2. Terry Weaving Flow Chart	23

	3.	Preparation for Weaving	24
	4.	Warping	24
	5.	Sizing	25
	6.	Main Sizing Agents	26
	7.	Sizing Auxiliary Substances	27
	8.	Drafting for Terry Weaving	27
	9.	Terry Designing	29
VI		Wet Processing and Finishing of Towels	31
		Product Design	31
VII		Physical Properties of a Terry Towel	37
	1.	Absorbency	37
	2.	Heat Insulation	37
	3.	Crease Resistance	37
	4.	Dullness	37
VIII		Advances in Machines for Production of Terry Towel Products	38
	1.	Stitching Automation-Monti Mac, Italy	38
	2.	Terry Towel Pretreatment & Dyeing-Toweltec by Fong's Hong Kong	38
	3.	Toweltec–2T High Temp High Pressure Soft Flow Dyeing Machine	38
	4.	Colour Kitchen–Eliar, Turkey	39
	5.	Rope Opener–Corino, Italy	39
	6.	Montexstenter–A. Monforts, Germany/Monfongs, Hong Kong	40
	7.	Weft Straightener/Pattern Control System–Mahlo, Germany	40
	8.	Continuous Bleaching/Pad Steam Dyeing/Washing Range–Goller, Hong Kong	41
	9.	Flatbed/Rotary/Digital Printing–Zimmer, Austria	41
	10.	Metal Detector–CEIA, Italy	43

IX		Quality Testing of Terry Towel and its Assurance	45
	1.	Introduction	45
	2.	Basic Parameters of Quality Terry Towel	47
	3.	Methods	47
	4.	Water Absorption Test	48
	5.	Dimensional Stability Test	48
	6.	Fastness Properties of Terry Towel	48
	7.	Fastness to Washing	48
	8.	Fastness to Light	49
	9.	Fastness to Rubbing	49
	10.	Always Check	50
X		Defects their Causes and Remedial Measures in Terry Fabric	51
		Introduction	51
		Problems Caused Due to Weft Breakages	52
		Formation of Cracks When Changing from Border to Pile	52
		Formation of Random Warp Wise Cracks	53
		Curled and Folded Surfaces	56
		References	58

CHAPTER I

INTRODUCTION

1. Introduction

A terry towel is described as a textile product which is made with loop pile on one or both sides generally covering the entire surface or forming strips, checks, or other patterns (with end hems or fringes and side hems or selvages). Terry fabrics basically belong to the group of pile fabrics, where in an additional yarn is introduced/inserted in such a manner that forms loop, called as pile, to give a distinct appearance[1].

Some 2.5 billion square meters of terry are produced in the world every year[3]. Demand of these products is increasing very fast globally and also significantly day by day in the local market. Consumers pay on an average $7 for bath towel and their households in America. In last decades, the importance of this sector has been increased enormously with the incremental global demand.

Table 1: Production and Shipment Figure of Terry Towel according to their End Uses (US) Census Bureau, 2005[2]

	2004		
Product Description	**Production**	**Shipments**	
	Quantity thousand dozen	**Quantity** thousand dozen	**Value** $1000
Finished towels	20,822	21,406	715,575
By end use			
Kitchen	(D)*	(D)	(D)
Bath	10,827	11,226	441,875
Hand ,face, guest, and fingertip	7,995	7,975	171,303
Bath/tub mats	(D)	(D)	(D)
All other	(D)	(D)	(D)
Wash cloths	13,300	12,724	111,882
*D The information was not disclosed to protect the privacy of the companies that produce these items			

This data which contain the volume of the towel market reveals the importance of the technology required for terry production.

2. History of Terry Weaving

The name "terry" comes from the French word "tirer" which means to pull out, referring to the pile loops which were pulled out by hand to make absorbent traditional Turkish toweling. Latin "vellus", meaning hair, has the derivation "velour", which is the toweling with cut loops.

In research conducted on terry weaving by Manchester institute of textiles, it was concluded that the original terry was likely a result of defective weaving. Terry weaving is considered a later development in evolution woven fabrics.

Victor Hobson (1990) described various mechanisms of terry loom being developed at initial stage. He described that at primitive stage, like other fabrics, handlooms were used to manufacture terry towels, but to produce loops, it was necessary to insert long rods in the same direction that weft were inserted. Length/size of loops was dependent on the thickness of rods. Following the invention of power operated looms, mechanical means were used to insert and withdraw the wires but these were no longer used in production of terry towels due to the complexity of operation. According to Hobson (1990), all the basic mechanisms for pile formation such as loose reed, shifting of cloth fell, variable fall back controlled by a pattern chain for sculptured effects, etc., had been developed by the end of 19th century.[1]

CHAPTER II

APPLICATION OF NATURAL FIBRES IN TERRY TOWEL MANUFACTURING

1. Introduction

Studies of woven fabric comfort properties have aroused the interest of researchers in recent years. Although studies on the structures of woven terry fabrics are rather limited, the study of the comfort properties these fabrics will reveal new approaches regarding the subject. Towels are the most used textile structures in water related usage of terry-woven fabrics. The users prefer that ready-made bathrobes and towels be comfortable and fresh, made of a light and soft structure, remain dry as they quickly transfer the water and sweat accumulated on the body, and be hygienic and naturally formed. Therefore comfort, an important property for the textile products, is also an important need for terry fabrics in water-related usage. However, the comfort properties of terry fabrics such as towels should be specific. The comfort parameters of air permeability, water vapor permeability, liquid transfer velocity, drying time, and water absorption will stand out in such products.

2. Cotton Fibers

Cotton fibres consist of the unicellular seed hairs of the bolls of the cotton plant, the Gossyum plant the chemical composition of typical cotton fiber is as follows: 94.0% of dry weight is cellulose, 1.3% is protein, 1.2% is pectic substance, 0.6% is wax, 1.2% is ash and 4% is other substances. Absorbency refers to a cotton fabric's ability to remove liquid water from the skin as in a towel. Cotton is hydrophilic; it wets easily, and can hold much more water than synthetic fibres can. Cotton releases a considerable amount of heat when absorbing moisture, but it dries slowly. It is not only the amount of water held that is most important, but the water held that is most important, but from the body. The size and distribution of the pores, and capillaries, between and within cotton fibres are uniquely suited for this purpose. Wet strength is one of the crucial properties required in towels, as they are most likely to remain wet as compared to other home textiles. Cotton is stable in water and its wet tenacity is higher than its dry tenacity. The toughness and initial modulus of cotton are lower compared to hemp fibres, whereas its flexibility and its elastic recovery are higher. Cotton is a natural fiber and considered hypoallergenic. This means cotton has a low tendency to cause allergic reactions. It also does not cause skin irritation and can be sterilized. The microbial resistance of cotton is low, but the fibres are highly resistant to moth and beetle damage. The microbial resistance can be improved by antimicrobial finishing. Cotton uses in the medical institutional area are well known for their hypoallergenic characteristic and sterilize- ability. Cotton fabrics are often recommended for persons having skin allergies. Cotton sanitary products and cosmetic aids are promoted for their health benefits. Cotton towels, bedding and baby clothes have all been promoted on the basis of the hypoallergenic nature of cotton.

Moreover cotton's resistance to high temperatures of water makes cotton easy to be cleaned as it can be boiled. Cotton fibres are the backbone of the Cotton fibres are the backbone of the it has the highest production and consumption figures among the other natural fibres. It has easy availability as it is grown in more than seventy countries of the world. One other reason cotton is used for toweling is it is the most economical fiber among the natural fibres Shorter staple cotton fibres are generally used in towels because fine yarn counts are not required. The cotton fibres which are used in towels have relatively low fiber length, relatively low fiber strength, relatively low maturity ratio. The micronaire range can be said to be in the middle range.

3.   Bamboo Fiber

Is a bamboo fiber as raw material, through careful design and multiple processing techniques to produce a set of health, environmental and aesthetic health in one of the new towels.

Has become a focus on health, the pursuit of quality of life, increase consumer choice bit of fashion. Bamboo fiber terry towel features:

1.   Bamboo fiber towel smooth, velvet has a unique sense of softness to the smooth skin of the most delicate care, skin care is doing my material of choice.

2.   Cross-section of bamboo fiber towels covered with large and small oval-shaped pores, the height of the natural hollow cross-section so that it can absorb and evaporate in an instant a lot of water, called 'Fiber Queen.'

3.   Bamboo contains a natural bactericidal components, it has antibacterial properties of mites. Textile products by the State Quality Supervision and Inspection Center for testing validation: the same number of bacteria under a microscope, bacteria in the cotton, wood fiber products can flourish, and bamboo fiber products, the bacteria killed in 24 hours after more than 80% antibacterial ability is unmatched by other textile materials.

4.   Bamboo fiber has been completely defatted, desugared, removal of protein processing, to prevent the stains on the towels in the bamboo fiber residue.

5. Bamboo fiber has been completely defatted, desugared, removal of protein processing, to prevent the stains in the bamboo fiber towel chemical reaction, has a strong cleaning ability, quick and thorough decontamination. Bamboo fiber is six ten thousandths UV transmittance, ultraviolet transmittance of cotton, two thousand five hundred ten thousandths, UV resistance of bamboo fiber is made of cotton 4 ten 7 times.

6. Compendium of Materia Medica' in the sterilization on the bamboo, clear and fire over 20 different medicinal functions and formulas of the elaborate, nearly a thousand species of bamboo folk prescription.

4. Hemp Fiber

Hemp fiber has been used extensively throughout history, with production climaxing soon after being introduced to the New World. Items ranging from rope, to fabrics, to industrial materials were made from hemp fiber. Hemp was often used to make sail canvas, and the word canvas derives from cannabis. Today, a modest hemp fabric industry exists, and hemp fibers can be used in clothing. Pure hemp has a texture similar to linen. Hemp fiber is one of the strongest and most durable of all natural textile fibers. Products made from hemp will outlast their competition by many years. Not only is hemp strong, but it also holds its shape, stretching less than any other natural fiber. This prevents hemp garments from stretching out or becoming distorted with use. Hemp may be known for its durability, but its comfort and style are second to none. The more hemp is used, the softer it gets. Hemp doesn't wear out, it wears in. Hemp is also naturally resistant to mold and ultraviolet light. Due to the porous nature of the fiber, hemp is more water absorbent, and will dye and retain its color better than any fabric including cotton. This porous nature allows hemp to "breathe," so that it is cool in warm weather. Furthermore, air which is trapped in the fibers is warmed by the body, making hemp garments naturally warm in cooler weather.

Hemp is an extremely fast growing crop, producing more fiber yield per acre than any other source. Hemp can produce 250% more fiber than cotton and 600% more fiber than flax using the same amount of land. The amount of land needed for obtaining equal yields of fiber place hemp at an advantage over other fibers. Hemp leaves the soil in excellent condition for any succeeding crop, especially when weeds may otherwise be troublesome. Where the ground permits, hemp's strong roots descend for three feet or more. The roots anchor and protect the soil from runoff, building and preserving topsoil and subsoil structures similar to those of forests. Moreover, hemp does not exhaust the soil. Hemp plants shed their leaves all through the growing season, adding rich organic matter to the topsoil and helping it retain moisture.

5. Wood Fiber

Generally speaking, high-quality wood fiber has the following features:

1. Soft and pleasant and will not harden. High-quality wood fiber is generally used in North America the growth of pine as raw material, through special refining process. The cork pine and growth characteristics of a longer period, determines the grade of cellulose wood fiber contained the highest number of directly created a super-soft properties of wood fiber, its textile products made of super soft and does not become The outstanding characteristics of hard, long-term use until the discarded remains soft and pleasant time.

2. Row oil decontamination, cleaning no trace.

3. Super moisture absorption, comfort, and body care.

4. Anti-bacterium, comprehensive odor.

5. Summer and autumn breathable, warm in winter and spring.

6. superior flexibility and lasting security type: high-quality wood fiber with superior flexibility, thus creating their product freely adjustable elastic, a pull-Yi Che that is able to easily put on or take off, and can long maintain its bright outside type.

7. Good drape, and never compacted. High-quality wood fiber has a natural drape, thereby creating products smoothness of its formation, long-term use of non-compacted, more close, beautiful and sexy.

8. High whiteness, spinning and strong.

9. Green environmental protection, natural health: high-quality wood fiber is a healthy type and environment-both natural plant cellulose fiber, textile products have a significant characteristics of healthy skin care.

CHAPTER III

PARTS AND CLASSIFICATION OF TERRY TOWELS

1. Parts of a Conventional Terry

A woven towel consists of five parts. These are the pile area, fringes, beginning and end part, selvedge, border. Every towel does not have to contain all of these parts. The pile area is considered the toweling part of the towel. Fringes are tied or an untied tasseled part of ground warps and pile warps which are left unwoven at the beginning and the end edges of the towel. The beginning and end sections are the tightly woven areas of a towel which come before or after the pile fabric part and prevent this pile area from unraveling. They are woven without pile loops, in a flat weave construction.

The selvedge contains fewer number of warp ends than the pile area, for example 90 comparing to 4000 total warp ends, woven without pile as a flat weave and has the purpose to reinforce the towel Sides.

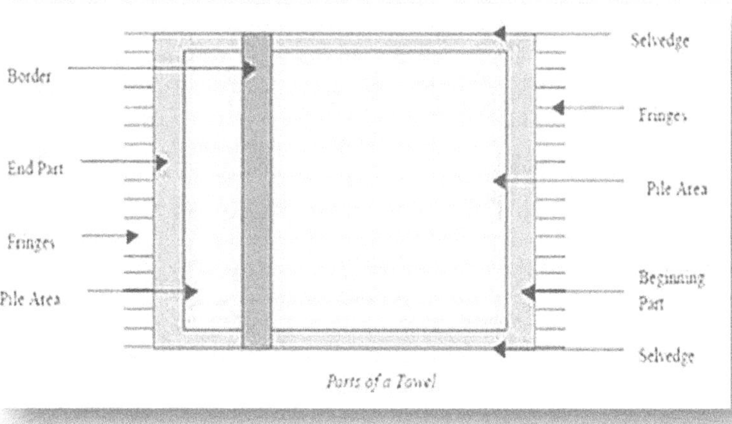

Figure 1: Different Parts of Terry Towel

2. Classification of Terry Towels

The classification of towels can be made according to weight, production, Pile presence on fabric Surfaces, pile formation, pile structure, and finishing. These classifications are shown in Table.

Table 2: Classification of Terry Towel According to Weight, Production Style, and Finishing Weft Count Per Pile Loop and Pile Presence on Fabric Surface

Weight	Production	Finishing	Weft Pick Count per Pile Loop	Pile Presence on Fabric Surface
Very Heavy (>=550 g/m²)	Woven	Velour Towel	Two-pick Terry	One Side Pile
Heavy (450-550 g/m²)	Weft Knitted	Printed Towel	Three Pick Terry	Both Side Pile
Medium (350-450 g/m²)	Warp Knitted	Towel with Embroidery	Four Pick Terry	
Light (250-350 g/m²)		Towel with Applique	Five –Pick; Six-Pick ; Seven or More Pick Terry	

CHAPTER IV

STRUCTURE OF TERRY TOWEL AND TERRY FABRIC WEAVING

1. Fibers Used in Terry Towels

According to Acar, the required properties of yarns which are used in terry towels are high absorbency, high wet strength, and ability to dye well, good color fastness wash-ability, soft hand, and hypoallergenic, low cost, and easy availability. Yarns made of cotton fibers can provide these properties most effectively.

Table 3: Ranges of Cotton Fiber Properties which are Used in Towel Fabrics according to US Cotton Fiber Chart

Fabric Type	Fibre Length in inch	Fibre strength in g/tex	Micronaire	Maturity Ratio
Toweling	0.93 – 1.10	20-32	3.5-4.9	0.80-0.90

2. Other Fibers

More and more towels are being produced from fibers other than cotton such as Modal®, bamboo, seaweed, Lyocel® and now soybean, corn and other Tri-blend bamboo, silk and cotton blend is also beginning to be used in towels. Bamboo may be the next premium fiber other than high quality cotton fibers because of its softness, luster, antibacterial properties and greater absorbency. Flax fiber has better dry strength than cotton, and like cotton it gets 25% stronger when wet. It absorbs more moisture, and it wicks. It is longer, smoother, and more lustrous than cotton. However it is not used commonly in towels as it has been limited in supply and it is expensive because of the long processing and intense labor it needs to be turned into a yarn although uncommon, flax towels have a place in the specialty market.

Micro-fiber towels are also pushing into the ultra-touch/high absorbency arena with a manmade synthetic product constructed primarily from a blend of polyester and nylon with polyamide. Like cotton, micro-fiber towels are available in various colors and weaves, such as waffle, cut terry and loop terry, with various patterns and in various weights. The heavier the micro-fiber towel, the more water it can absorb.

3. Yarns which are Used in Terry Towels

In a terry towel there are four groups of yarn. These four groups are the pile warp, ground warp, weft (filling), and border weft.

3.1. Pile Warp

One hundred percent cotton yarns, carded or combed, in sizes of 16/1, 20/1 Ne counts, 240-255 turns/meter twist, are most commonly used. The use of cotton-rayon blends has diminished, because100% cotton provides a more pleasing hand and texture then the blends. When high quality is required, two or more ply yarns are used. In this case absorbency increases, and the fabric gains resistance to pile lay. The use of two-ply yarns is also on the increase as it improves visual appearance. Plied yarns are used to form upright loops in classic terry, whereas single yarns are used to form spiral loops in fashion terry known as milled or fulled goods. The two types of loops are shown.

(I) is an upright loop and (II) is a spiral loop

Figure 2: Two Types of Loops

In the first type of classic terry patterns are usually created by employing dyed yarns; while towels of the fashion type are mainly piece dyed or printed. In general bulkier and absorbent yarns are used for both types of towels. In real Turkish-toweling, the pile-loops generally consists of a more highly twisted yarn which, while very absorbent, are quite abrasive, thus actively stimulating the skin during drying. Rotor spun yarns are also used in pile warps low twist cotton.

3.2. Ground Warp

Carded yarns of 20/2, or 24/2 Ne count with 550 turns/meter twist, and of 100% cotton are commonly used for ground warp ends. Two ply yarns are preferred because the ground warps ends have the highest tension during weaving. It is common to use a yarn of cotton/ polyester blend for greater strength. Rotor spun yarns are also used in ground warps.

3.3. Weft

Carded yarns of 16/1, or 20/1 Ne counts with 240–255 turns/meter twist, 100% cotton are used usually for weft or filling picks. Rotor spun yarns are also used in wefts.

3.4. Border Weft

Premium or high end hand towels have complex borders with fancy weaves and use a very wide range of filling yarns. Decorative, shiny and bulky yarns of rayon, viscose, polyester, chenille, or mercerized cotton are used at different yarn sizes. Novelty types of yarns may be used as a feature of design

4. Construction of Terry Towels

Terry towels are woven as 2, 3, 4, 5 or more pick terry weaves. The most common type is 3-pickterry toweling. The cross section of a toweling through the Warps are divided into two systems as shown in figure , pile warps and ground warps, whereas wefts consist of only one system.

In basic Turkish Toweling, front side and back side pile warps and 1st and 2nd ground warp ends form a 2/1rib weave with each other. The rib weaves which is formed by the pile warps is one pick ahead of the rib weave which is formed by ground warp ends. Warps are ordered throughout the fabric width 1:1or 2:2 piles and ground warps. In 1:1 warp order each ground warp end is followed by a pile warp end while in 2:2 warp order each two ground warp ends are followed by two pile warp ends. The weave notation of 3 weft pile basic Turkish toweling is given in 1:1 and 2:2warp orders.

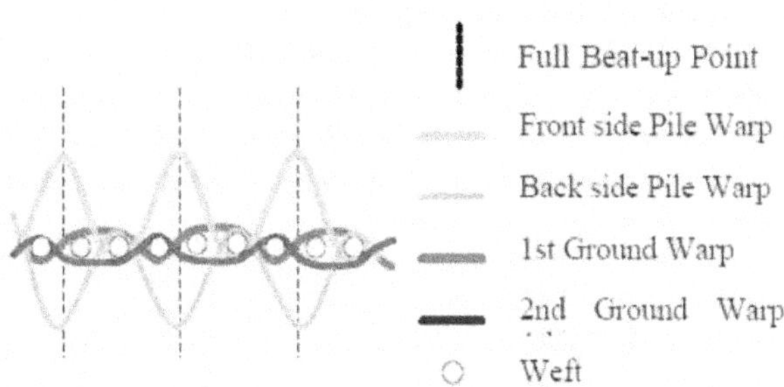

Figure 3: The Cross-section of Towel through the Warp

As is seen from the weave diagrams in figures a and b, the shedding of the ground warps are not synchronized with that of the pile warps. By this, the number of interlacing throughout the warping creases and this strengthens the fabric. As it has been mentioned before terry towels can have pile loops on one or both faces. Different types of terry weave which have pile on one face and both faces.

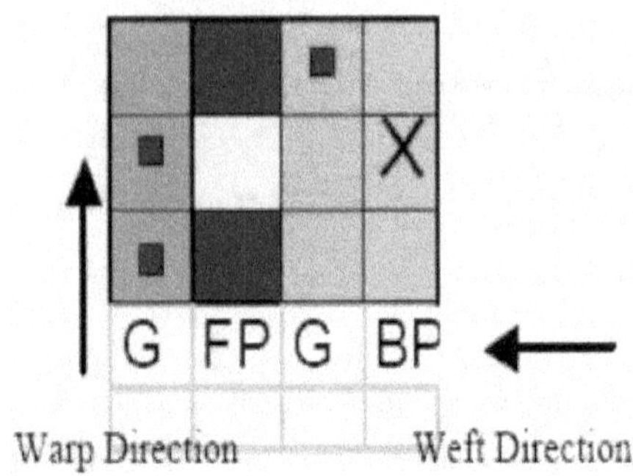

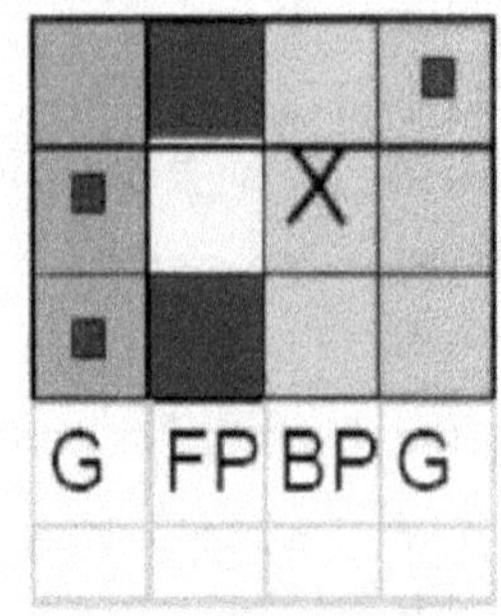

a. Basic 3-pick Terry Wave in 1:1 Warp Order, b. Basic 3-Pick Terry Wave in 2:2 Warp Order

Figure 4: Basic 3-pick Terry Weaves with Different Warp Order

G: Ground Warp

FP: Front Face Pile Warp

BP: Back Face Pile Warp

Little block: Ground warp is over the weft

Shaded: Front Face Pile Warp is raised over the weft

X: Back Face Pile Warp is raised over the weft

Empty space: Warp is lowered behind the weft

The weft count used for toweling is between 15 and 25 picks/cm. And warp count is between 20 and30 ends/cm. During the weaving of borders, the weft count is increased 3 to 6 times the density in the pile areas Pile/ground ratio is described as the length of pile warp per unit length of fabric in the warp direction. A practical way to find out this ratio is done by measuring a 10 cm length of toweling in the warp direction, then cut the pile warp from either ends of the measured length and measure the total length of the removed pile end per 10 cm length of fabric.

Pile warp length per 10 cm fabric is usually, between 20 to 100 cm and his ratio has a direct effect on the fabric weight and thickness. As the ratio increases, the weight and the thickness of the terry fabric increases.

Pile warp yarns constitute the largest portion of terry fabric weight. The pile warp constitutes between 65.0% and 79.0% of the total weight, depending on the terry fabric construction and pile length.[4]

5. Loop Formation Technology

In the production of terry fabrics two warps are processed simultaneously: the ground warp, with tautly tensioned ends, and the pile warp, with lightly tensioned ends.

A special weaving method enables loops to be formed with the lightly tensioned warp ends on the fabric surface. With the basic method, known as three-pick terry (Fig. 5), three picks form a pick group. By means of a special device on the weaving machine, two picks are inserted at a variable distance–the loose pick distance–from the cloth fell. The loose pick distance is varied according to the desired loop height. When the third pick is beaten up, the reed pushes the pick group on the tautly tensioned ground warps towards the fell, and the loose pile warp ends woven into the pick group are up righted and form loops.

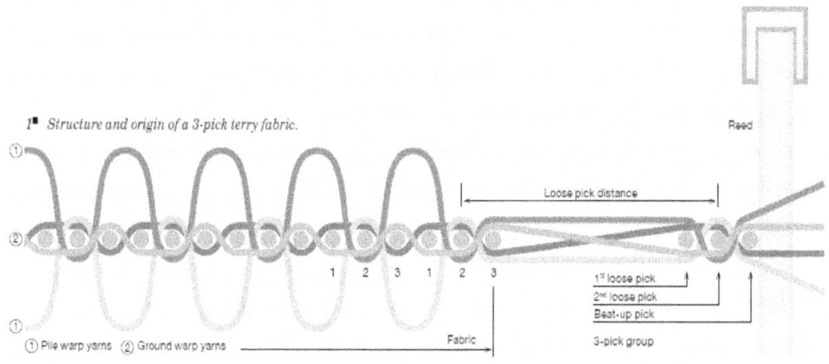

Figure 5: Three Pick Terry

If piles are to be formed on the surface of a terry fabric, the pile warp yarns must be over the third and first picks; similarly, if the piles are to formed on the back side of a terry fabric, then the pile warps must be under the third and first picks.

6. Steps of Terry Weaving

The components of an air- jet terry weaving machine are seen. The pile warp ends are let off from the pile warp beam (2), guided through the measuring unit (3), then join with ground warp ends which are let off from ground warp beam (1) and guided through the whip roll. Next, the two warp systems are threaded through the drop wires, the headless, reed and with the control of cloth take up (6) are wound onto cloth roll after weaving(7). Positive controlled whip roll for ground warp (5) determines the length of ground warp to be let off, while terry motion (4) assures integration among pile and ground warp let off and cloth take up.

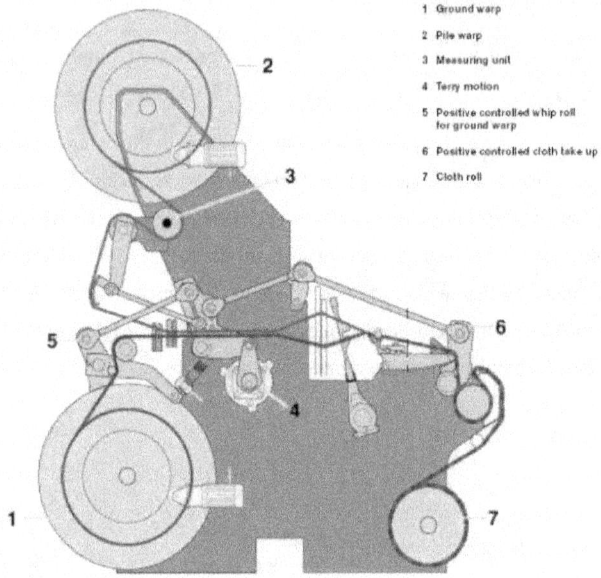

Figure 6: Components of Air Jet Terry Weaving Machine (Dornier 2003)

7. Basic Movements

Pile warp threads form loops and patterns through the shedding motion where as the ground warps form the ground with 1/1 plain weave, rib 2/1, rib 2/2, or rib 3/1 weaves. The rib 2/1 weave is the most frequently used weave for pile and ground warp systems separately. The shedding motion can be controlled in three ways for terry weaving as follows.

8. The Shedding Mechanism

8.1. Cam Shedding

The shedding motion is applied to the warp threads through heald frames or harnesses. The maximum number of frames which is used in terry weaving machines is 10. This system can weave only very basic weaves. Weaving machines can reach very high speeds with an eccentric shedding system. To change the weave it is necessary to change the cam.

8.2. Dobby Shedding System

The dobby shedding motion of the warp ends is created by the movement of the heald frames. The maximum practical number of harness frames in terry weaving machines with dobby is 20. Dobby looms can produce weaves in limited numbers and limited designs. The difference of this system from the other traditional dobby looms is that the motion of the pile and ground warps is transferred separately.

There are also systems in which the pile warps are given shedding motion by dobby and the ground warp ends by cam.

8.3. Jacquard Shedding System

Each warp end is controlled through a separate motion. Very different and very complex structures can be woven. In terry fabric it is the pile warp which shows the design. The ground warp ends are woven 2/1 ribs (most commonly), 2/2 ribs, 1/1 plain weave and 3/3 ribs (the rarest) As these weaves do not require a jacquard shedding system, in some weaving machines, the pile warp ends take the shedding motion from the jacquard system whereas the ground warp ends take from the cam.

Jacquard machines may either work in a traditional mechanical system with the help of needles and design cards or a contemporary electronic system which works through electronic transmitting elements with design files and electronic. The pile warp ends are looser than input. The pile warp ends are looser than ground pile ends, thus the shedding of the pile weft ends must be wider than that of ground warps. Otherwise, contact may occur between the pile warp ends and the pick carrying device, which may cause high numbers of end breakages. A wider shed also improves the loop formation. Pile tension rod should guide the pile warp ends from the position which has the same level with the center of the shed. With this, the pile warp ends in the upper and in the lower shed will maintain the same tension. Thus, the pile loops on both sides of the fabric will have the same length.

9. Filling Insertion

9.1. Filling Insertion with Rapiers

Rapiers are popular in the production of terry cloth because of the flexibility they offer for production.

Rapiers are two hooks which carry the weft picks across the warp sheet. The first giver hook takes the weft pick from the yarn feeder and carries it to the center of the warp width. Meanwhile the taker hook moves from the other side of the weaving machine to the center. There, the two hooks meet and the weft pick is transferred to the taker hook. After that the giver hooks returns empty to the side it came from, and the taker hook carries the weft to the opposite side.

9.2. Filling Insertion with Air Jet

In air jet weaving a puff of compressed air carries the weft yarn across the warp sheet; there are relay nozzles which are arranged in a definite order according to fabric width. These aide nozzles are connected to the main nozzles in groups. The air hoses which go to aide nozzles are also arranged in a row. The pick feeders also work with air and winds according to the fabric width. On the side where the pick arrives there are optical sensors which control the arrival of the filling picks. The maximum filling insertion rate practically achieved in terry weaving is 1800 m/min

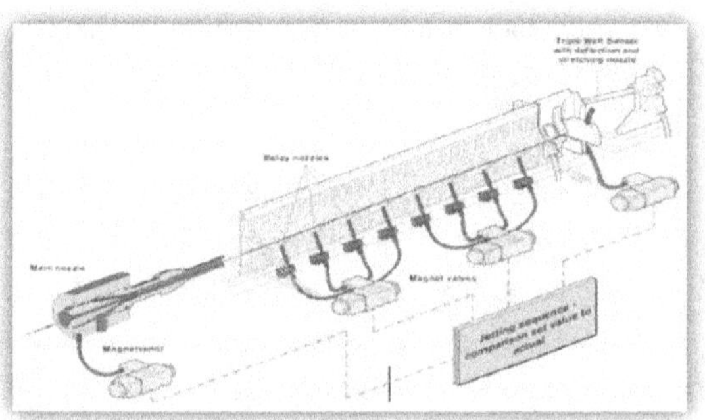

Figure 7: Air Filling Insertion System

9.3. Filling Insertion with Projectile

A small gripper takes the cut weft yarn across the weaving loom. This system is not very common in terry weaving as rapier and air jet filling insertion system are most commonly used ones Promatech,2003; Dornier, 2003 Picanol, 2004; Smit Textile, 2005 Tsudakoma, 2005.

10.Beat-up

The loops in terry fabrics are formed with a special reed motion and warp let- off system. These motions vary according to pick number per loop. In 3-pick terry weaving, two picks are inserted at a variable distance the loose pick distance-from the cloth fell. The loose pick distance is varied according to the desired loop height. When the third pick is beaten up, the reed pushes the pick group which includes the three picks, on the tightly tensioned ground warps, towards the fell and the loose pile warps are woven into the pick group are up righted and form loops. Depending on the weave, loops are thus formed on one or both sides of the fabric.

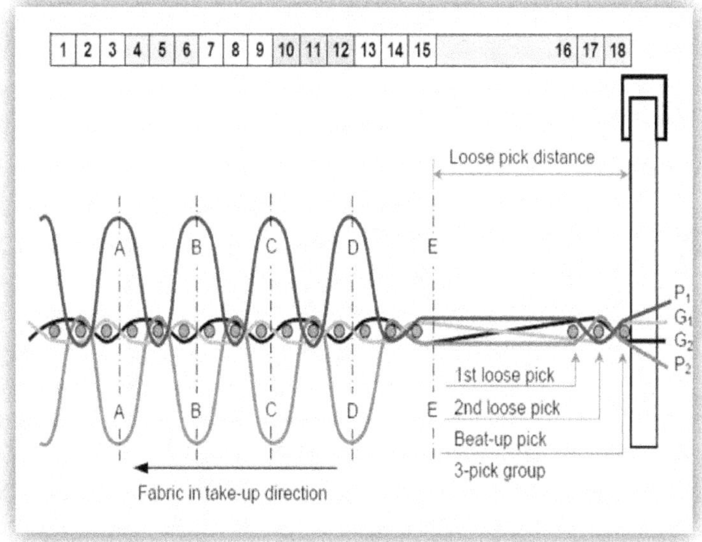

Figure 8: Three Pick Terry Fabric Formation

In Figure above, 3-pick terry fabric formation is seen. The first weft pick is the loop fixing pick, the second pick is binding pick, and the third one is the pile pick or the fast pick. The third pick is inserted into a completely reversed shed, as the pile and ground warp ends which are up, go down, and those which are down go upward, essentially locking the first in place. Thus, this motion prevents the drawing of the loop by the following sheds.

There are also systems in which the reed motion is constant but the cloth fell is moving, like Zax-e® Terry loom from Tsudakoma which has terry motion with a cloth fell shifting system or the ATVF ServorTerry® weaving machine from Dornier (Tsudakoma,2005; Seyam, 2004). Here, a servo motor replaces the traditional terry cam for pile formation, so the reed does not drop back. When the reed is at the front center the fabric is positively driven toward the reed to form pile by the backrest and terry bar in combination with the temples. The disadvantage of this system is that the friction which takes place during the forward-backward motion of the ends can lead to end breakage. Although weaving machines of different makes have different mechanism the main principle is the same. With today's machines, the maximum loose pick distance practically achieved is 24 mm, which gives some less than 12 mm loop height in G6300F® Terry Weaving Machine. It is possible to switch between 3-, 4-, 5-, 6- or 7-pick terry and 8 different pile heights in ServoTerry® (Dornier, 2003) and G6300F Terry Weaving Machines while the machine is running a towel which is woven with different pile heights is seen.

11. Complementary Motions

11.1 Let-off

It was mentioned earlier that there are two warp systems including ground warp and pile warp, and thus two warp beams are let off simultaneously in a terry weaving machine. The ground warp ends move forward slowly and under high tension as the ground warp beam turns slowly. At the same time, the pile warp ends move forward quickly and loosely as the pile warp beam turns faster than the ground warp beam. Ground and pile warp beams are propelled by two different independent motors.

Rpm's (revolution per minute) of the pile warp beams is proportional to the required pile height. The higher speed delivers more yarn to increase the pile height. During let- off, pile tension is controlled continuously. This decreases yarn breakages, and avoids out-of tolerance loop heights. In the following figure the Terry Motion Control System® of Tsudakoma is shown. Here, pile tension is determined by pile tension roll which is propelled by a motor guided by electronic pile tension control system allows, so that it can hold the maximum length of pile warp. Keeping the pile beam's diameter large avoids changing the beam frequently.

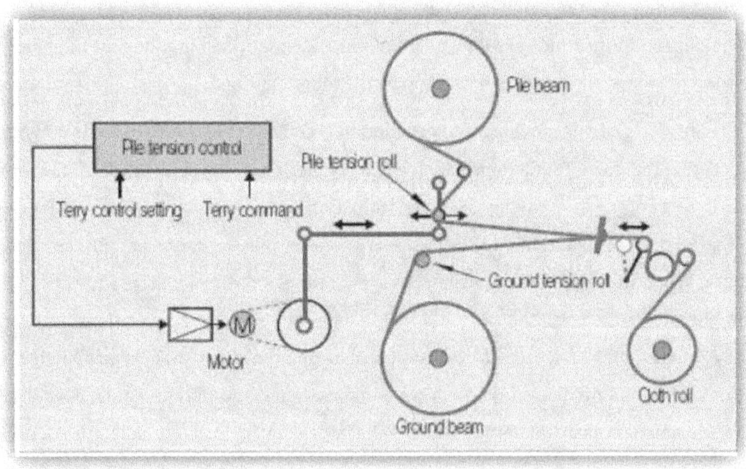

Figure 9: Pile Tension Control System (tsudakoma)

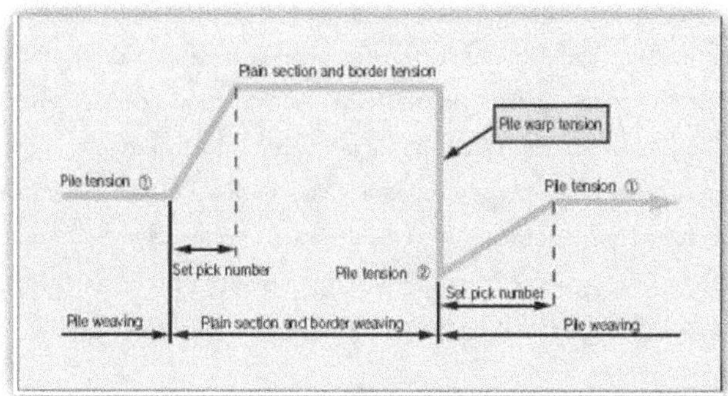

Figure 10: Diagram of Pile Warp Tension during Weaving, Pile, Plain and Border Parts (tsudakoma)

The width of the pile beam is between 76–144 inches (190-360 cm) and the diameter of its flange can be up to 50 inches (125 cm), while the flange diameter of the ground beam is up to 40inches (100 cm). The Pile beam can hold more than 130 cu ft of yarn, with a gross weight exceeding that of many automobiles.

The two warp systems are evenly let-off by a system of constant tension control from full to empty beam. This is controlled by a highly sensitive electronic device. The tensions of the pile and ground warps are detected by force sensors and electronically regulated.

Elimination of unwanted increase of tension of warp tension during weaving high density border and/or plain section is achieved by reducing let-off speed.

The diagram of pile warp tension in Zax-e® Terry looms from Tsudakoma during weaving of pile, plain and border areas is shown above. In the following Figure the diagram of loom rpm's in Zax-e® Terry looms of Tsudakoma during pile weaving and border weaving is shown

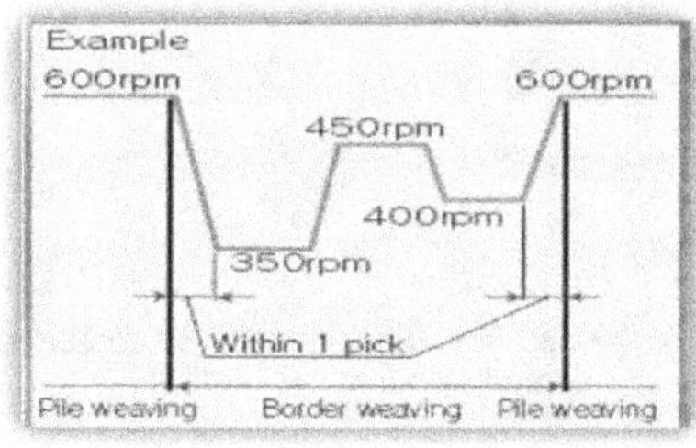

Figure 11: Diagram of Loom Rpm`s during Weaving Pile and Border Areas

To prevent starting marks or pulling back of the pile loops, the pile warp tension can be reduced during machine standstill. An automatic increase in tension can be programmed for weaving borders to achieve more compact weave construction in order to ensure a rigid border and/or to achieve nice visual effects via jacquard or dobby designs on the border. The way the back rest roller system is controlled depends on the weave. During insertion of the loose picks and during border or plain weaving the warp tension between the open and closed shed is compensated for by negative control. A warp tensioner with torsion bar is used for the ground warp, and a special tension compensating roll is used for the pile warp.

11.2 Take-up

The pick density is automatically controlled by synchronizing the take-up motor rotation with the loom speed. The take-up motor rotates the cloth pulling axle. The cloth pulling axle is covered with needles which pricks the terry fabric and assures that the thick fabric winds on the take-up roll evenly with a constant width. The electronically controlled cloth take-up guarantees exact weft densities in every terry towel and a faultless transition between pile and border. There are five elements of a take-up system.

These are:

11.2.1 Temple

The temple holds the width of the fabric as it is woven in front of the reed and assures the fabric to be firm at full width. A temple is seen on Figure.

11.2.2 Length Temple

Length temple is located on the center of loom width between two side temples. There are groves starting from the center and going to the left and right sides of the temple. It ensures the terry fabric is open to the sides and remains straight and tense throughout the fabric width.

11.2.3 Cloth Pulling Axle with Needles

It ensures the thick terry fabric keep its tension and width while being transferred from the length temple to the cloth transfer axle.

11.2.4 Cloth Transfer Axle

It increases the contact angle between the terry fabric and cloth pulling axle with needles and transfers the fabric to take-up roll.

11.2.5 Take-up Roll

The fabric which comes from the transfer axle is wound on take-up roll.

12. Auxiliary Motions

12.1 Selvedge Forming

A length-wise edge of a woven fabric is called selvedge or selvage. The main purpose of the selvedge is to ensure that the edge of fabric will not tear when the cloth is undergoing the stresses and strains of the finishing process. This is achieved by making the selvedge area stronger than the body of the cloth using heavier and plied warp yarns, increasing warp yarns per inch, and applying different weaves. Two types of selvedge are formed during terry weaving.

12.2 Leno Selvedge

A leno weave at the edges of the fabric locks in the warp yarns by twisting the last two warp yarns back and forth around each pick. They are made with special leno weaving harnesses. Leno selvedges predominate in terry weaving.

12.3 Tuck-in Selvedge

The fringed edges of the filling yarns are woven back into the body of the fabric using a special tuck-in device. As a result the filling density is doubled in the selvedge area.As the width of the towels is usually much narrower than that of the weaving machine width, more than one towel may be woven at the same time. Thus, selvedges are formed not only at the sides but also several selvedges should be formed on the sides of each towel panels woven together. For this reason special selvedge forming systems are produced for terry weaving. One example is Dornier's Pneuma Tuckers® for outside and center selvages, which are the selvedges of individual towel panels when they are woven on a loom side by side.Tucked in selvedge reduces the finishing costs of towels.

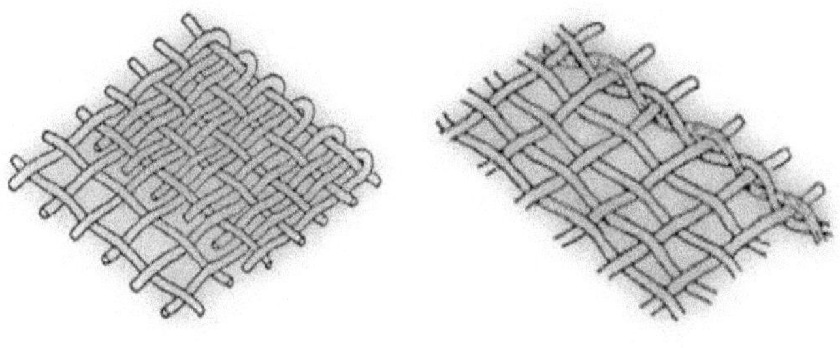

I) Leno Selvedge   II) Tuck in Selvedge

Figure 12: Common Types of Selvedges on Terry

12.4 Weft Color Choosing Motion

There are special color selection systems for inserting the required pick color while weaving different filling colors. Terry weaving machines have weft maximum twelve different colors or type of filling to be woven, including novelty yarns like chenille.

12.5 Pick Control

The pick control mechanism or pick finder detects the weft breakage. At a filling break, the machine stops and moves at reverse slow motion–automatically to free the broken pick. It has a significant role in reducing the down times for repairing filling breaks and thus the starting marks can be avoided.

12.6 End Control

Drop wires which are hung individually on each warp end, fall down when a warp end is broken or is very loose, closes down the electric circuit and thus shutting down the weaving machine.

12.7 Weft Measuring and Feeding Motion

During terry weaving in shuttle–less looms, the weft is inserted from one side with the help of rapiers, or air jet nozzles. A predetermined length of weft yarn under the necessary tension should be inserted during each picking. Before each picking motion, a definite length of weft pick is measured and, stored usually on drum accumulators and released for picking. The weft feeders carry out this function. They pull the weft picks from the yarn packages and wind them helically over a turning cylinder. Winding speed determines the weft length.

CHAPTER V

TECHNOLOGY OF TERRY TOWEL PRODUCTION

Terry towel production processes include spinning, weaving, dyeing and finishing, and cutting as general steps. Shearing and embroidery are also regarded as necessary sub steps to obtain the final product of a terry woven towel.

1. Terry Weaving

The production of terry fabrics is a complex process and is only possible on specially equipped weaving machines. Three yarn systems are woven in the terry loom compared to the two system types of traditional weaving: Ground warp, pile warp and weft. The two warps are processed simultaneously: the ground warp, with tightly tensioned ends and the pile warp with lightly tensioned ends. A special weaving method enables loops to be handled with the lightly tensioned warp ends on the surface.

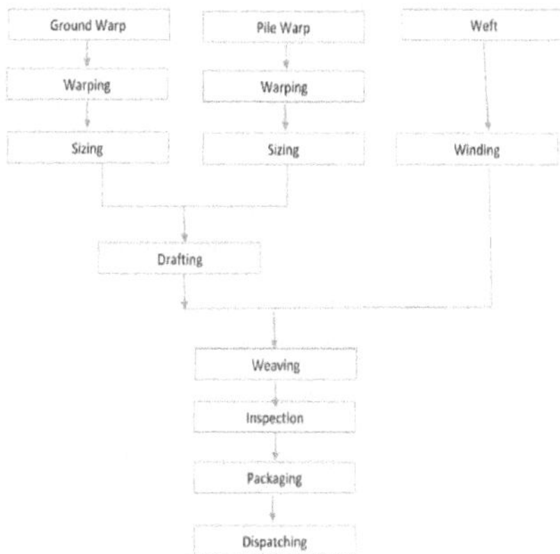

2. Terry Weaving Flow Chart

Ground warps and pile warps are unwound separately, warped onto two different section beams and sized separately.

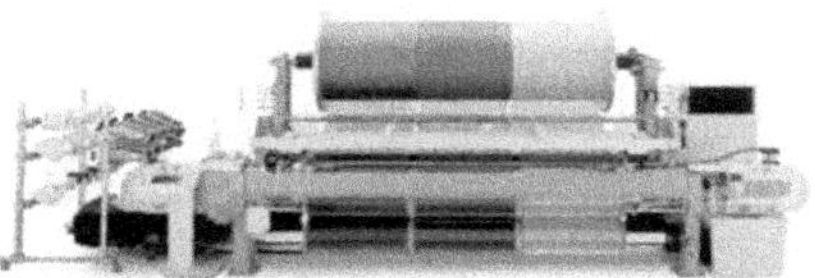

Figure 13: Terry Loom with Two Beams

3. Preparation for Weaving

Increasing demands are being made on warp quality due to the ever increasing speed of looms and weaving machines. Weaving preparation consists of procedures which are carried out before weaving in order to obtain good quality fabric by ensuring warp and weft performance Differences from each other.

Weft or filling yarns are wound onto bobbins in required softness and lengths. The flow chart of terry weaving process is in drafting or drawing in, ground and pile warps are passed through heddle eyes in the healed frames or harnesses, through ground and pile drop wires and through special terry reeds which have double teeth. Warps are fed into the loom from two beams: The ground and pile warp beams. The tension of the pile warp beam is lower than that of the ground warp beam; therefore the pile warp beam delivers higher length of warps than does the ground warp beam does. A special reed motion lets this extra length of pile warp form loops. Terry weaving is described as "slack tension warp method"

4. Warping

Warp ends should be wound onto the section beam in accordance with the required weave, total number of ends, length and the required warp density (epi) of the fabric. By setting the yarn tension consistently during warping throughout the warp beam, the sizing may be applied in a more homogenous manner throughout in a more homogenous manner throughout the beam. The objective of the warping systems is to present a continuous length of yarn to the succeeding process with all the ends continuously present and with the integrity and elasticity of the yarn as wound; fully preserved. In this process, yarn ends from packages which are placed on the warping creel according to the specified warping plan are wound onto the beam after passing through guides, tension regulators and the accordion comb. Two systems can be used for the warping process: Direct warping and sectional warping. If the creel capacity is sufficient and the number of total warp ends is not very high, the ends which are drawn from the creel can be wound directly onto the warp beam or section beam. This system is known as the direct warping system. If the fabric width is high or the warp density is so high that it necessitates a high number of warp ends or the warp has a color repeat, warping is carried out section by section. In this system which is known as sectional warping a definite number of warp ends are unwound from the creel and are wound on a cone shaped warping drum forming a specified width. The process is repeated until the required end count is reached. In the second step the warp ends which are wound onto the cone drum are transferred to warp beam. In direct warping the warp ends are wound onto a number of beams which will be joined in one weaver's beam after sizing, whereas in sectional drawing all the warp ends can be wound onto a single beam. Direct warping is much faster and thus cheaper than sectional warping as it includes only ones step for the warp ends to be wound on to the warper's beams During warping, the zigzag comb moves upward and downward to the left and to the right as the warp flows in order to prevent warp ends mounting one over other. The warp ends which come through the zigzag comb are wound onto the warp beam with the help of a transferring drum. The pressure drum ensures the tightness of the sizing beam and consistent tension across width and length.

The running speed of the warper and the tension of the ends can be increased as the thickness of the yarn increases. The speed of the warper is also affected by the type, the strength, and the friction of the yarn. Also, the speed of the direct warper is higher than that of sectional warper Yarn packages for warp beams are placed on the warping creel, either V- or parallel creel. A parallel creel has the advantage of space saving on the plant floor and the V–creel has the advantage that the tension is kept constant throughout the beam width Packages are arranged on the creel according to the color pattern repeat. Each warp end is tied to the end which was left from the previous beaming work and is passed through the tension regulator. The packages should be in good condition and all should have the same weight. Packages of different weight run out of yarn at different times: thus they would need to be replaced at different times. This leads to loss of production time. Moreover the yarns left on the package after the work is done are transferred to another to either form a full package or be sold off at discount prices as yarn waste. This leads to higher costs and loss of profit.

Each one of the beams in a set which will be sized is required to contain the same length of warp ends. Yarn breakages can occur due to thin places, nappy yarn, fly (the flying fibers from the yarns), bad package winding, or insufficient twist. If the creel is equipped with an air blower, the breakages which are due to fly will be prevented. The zigzag comb also should be equipped with an air blower.

5. Sizing

Terry toweling is formed from cotton yarns, and as described earlier these yarns are produced by gathering cotton fibers together and twisting them. Some of the fibers in the yarn are totally in contact with other fibers, while some fibers are loose and protruding. Fibers of the latter type do not contribute to the strength of yarn totally and form a rough yarn surface. Warp ends should be able to withstand great tension and friction forces during shedding and beat-up in the weaving process. As the number of end breakages increase so will the total cost and number of fabric defects increase Sizing is a pretreatment for yarns to be processed as warps for weaving into textile fabrics. Sizing protects the yarns against mechanical stresses in the weaving process by the application of a film of sizing agent which envelopes the yarn and which subsequently must be removed in finishing. The composition and quantity of the size application must be adapted to the type of yarn. Weaving efficiency is highly dependent on sizing. The type of sizing agent is also important in finishing which is why close cooperation between the weaver and the finisher is desirable. Sizing is carried out in warp form from beam to beam. For the warp sizing process, the size has to be cooked in a kettle after which the size liquor is transferred to a heated storage vessel. The liquor is delivered from the vessel to one or several size boxes for application to the warp sheet. The sizing machine can have one size box or more than one size boxes to increase the effect of sizing. The warp is squeezed between one or several pairs of rollers to remove excess size and to improve size penetration into the yarns. The impregnated warp then passes over drying cylinders supplied with heated steam. During this stage, water evaporates from the wet yarns and is normally collected under a hood and discharged by means of an extractor fan into the atmosphere through the roof. A sizing machine is Sizing liquor consists of three main components, main sizing agents, auxiliary sizing substances and water.

6. Main Sizing Agents

The sizing agents which are used today can be either natural sizing agents (starches, starch derivatives, cellulosic sizing agents) or manmade sizing agents (polyvinyl alcohol, acrylic) The most frequently used natural sizing agent is a starch derivative, carboxymethyl cellulose, which forms relatively more elastic but less strong size as compared to other sizing agents. The manmade sizing agent which has a widespread use is polyvinyl alcohol (PVA). The viscosity of the liquor can be adjusted during production. Film strength is high but its sticking ability is a little low. It dissolves in water. The desizing process should be carried out carefully; otherwise problems can occur in the wet agent is acrylic. It is usually used with other sizing agents to improve sticking ability and to prevent the size from settling Terry toweling is a heavy fabric and most of this weight comes from the pile warp. This situation increases size consumption and consequently increases sizing and desizing costs. An industrial application is to use starch or carboxymethyl starch in the ground warp, and carboxymethyl starch which can be removed in water for the pile warp. An industrial application is giving pile warp 3.5-4% size add-on and ground warp 13-14%.

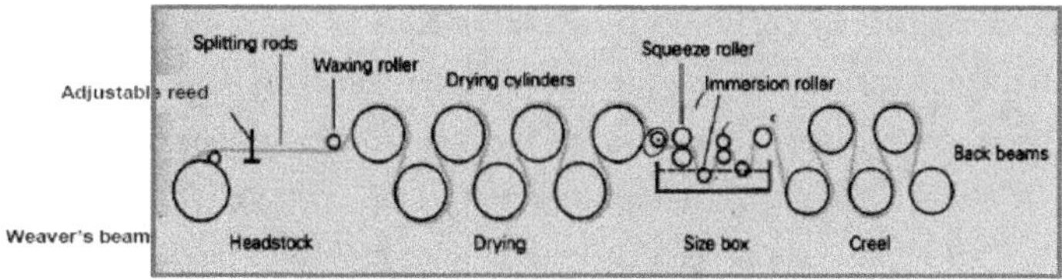

Figure 14: Sizing Machine Pictures

7. Sizing Auxiliary Substances

Sizing auxiliary substances are tensids, which help yarn absorb the liquor; softeners, which are used in order to soften the size film; lubricators, which are used to decrease the friction coefficient of sizing film, increase elasticity, improve moisture absorption; anti-static agents, which are used to prevent static electric; moisture holders, which are used for sizing film to ensure they hold 7-8.5% moisture until the end of weaving; de- foaming agents and antiseptics.

8. Drafting for Terry Weaving

Drafting or drawing-in is the process of passing the warp ends through drop wires, heald or heddle eyes and reed dents in the designated order. With this step the warp ends are arranged in the required order, prevented from crossing over each other, and warp density is set. It is one of the most laborious of all textile processes, however, most weaving mills throughout the world continues to do this process by hand. In terry weaving two ends are drawn through each dent. The reed numbers which are most commonly used in terry weaving are 110/2, 115/2, 120/2. Here the first number gives the number of the dents on the reed per 10 cm, and the second number gives the number of warp ends which pass through one dent. The reed which is used for terry weaving is different from that of normal weaving. The distinguishing characteristic of this reed is that its dents are arranged in two rows. This double row prevents entanglement of pile and ground warp ends, but this has a disadvantage. Any reed mark on the fabric becomes more obvious. However this makes it easy to distinguish the weave from the 3- or 4- pick terry fabric. In Figure 15, a reed which is used for terry weaving.

Figure 15: Reed Used for Terry Production

During weaving the flow of the warp ends should have as few obstacles as possible. Thus straight drafting is applied for pile warp ends. Straight drafting is achieved as the first end through first harness, second end through the second harness, to the final number of harnesses. The order is sequential. If the warp density is high, skipping drafting can be used for both pile and ground ends. In skipping draft the drafting order does not follow the sequential order of ends.

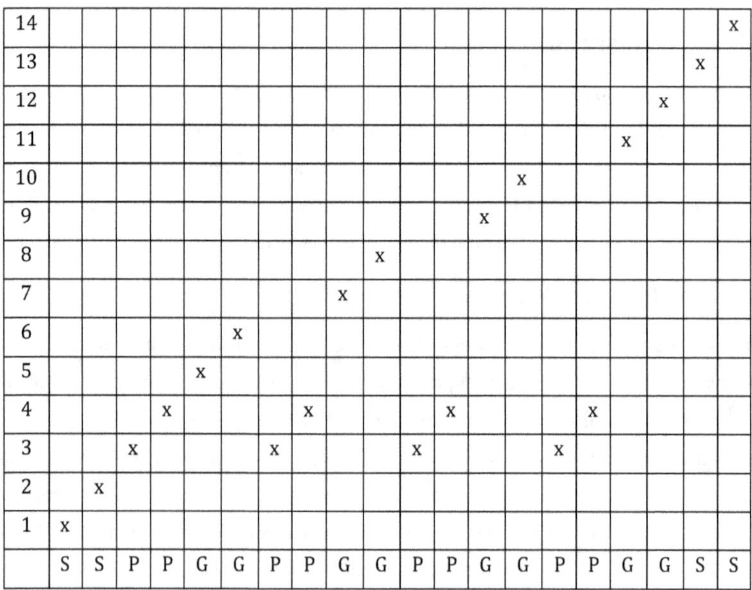

Figure 16: Drafting in Terry Weaving

P: Pile Warp ends

G: Ground Warp Ends

S: Warp ends of Selvedges

Leno selvedge warp ends through heald frames or harnesses 1 and 2

Pile warp ends through heald frames 3 and 4

Ground warp ends through heald frames 5 to 12

Leno selvedge warp ends again through heald frames 13 and 14

Pile warp adds only use two heald frames whereas ground warp threads which have a very close number of ends use 8 heald frames.

9. Terry Designing

Terry fabrics are often very complex with different colored warp ends in combination with loop patterns. They are subject to changing fashions, and the market is constantly demanding new qualities and designs. The rapid development of electronics has enabled fabric designers to produce completely different patterns. Via a servo motor, the beat-up position for each pick, and, thus the type of terry and the pile height can be freely programmed from one pick group to another. In this way nearly 200 different loose pick distances, and hence the same number of pile heights, can be programmed in any order. For example, three and four-pick terry and even fancy types of terry can be combined in the same fabric. This gives the fabric designer a broad range of patterning options and the weaving engineer the weaving structure for improving fabric performance, because transition from one pattern element to the next can be woven with greater precision With these capabilities, a new patterning method, called sculptured terry, has been developed. At each full beat-up, two pile loops of different heights can be formed in the filling direction. The secret of this method of pattern formation lies in the fact that two loose pick groups formed at distances corresponding to the pile heights are beaten up to the cloth fell together.

For two short loops the pile yarns are woven into both loose pick groups and for one large loop into the second loose pick group only. The greatest challenge is to develop a basic weave which results in neat loops without excessive friction between warp and filling at full beat -up. The solution is found in a special seven pick weave combined with full beat -ups at the sixth and seventh pick. In this way, a second pile height is also formed in filling direction, making sculptured patterning possible by the difference in pile height in warp and filling direction.

In the following a terry towel pattern which is produced with this technique is shown.

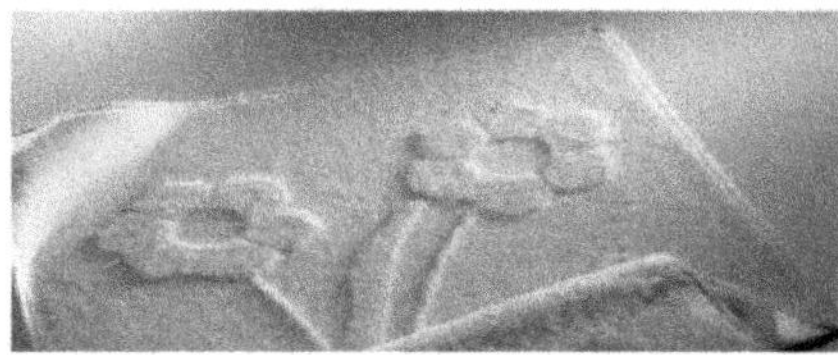

Figure 17: A Terry Pattern Achieved by Weaving Two Different Heights of Loops

In the following figure the diagram of seven pick terry design is shown. A requirement for this kind of pattern formation is a freely programmable sley traveling on a rapier weaving machine. Microprocessor control allows the loose pick distance to be programmed easily and individually for each pick.

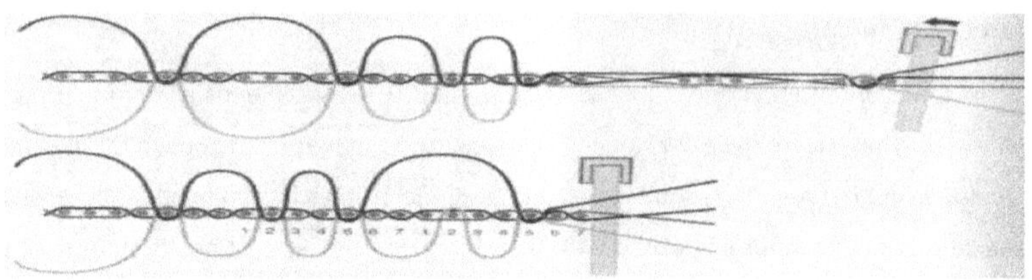

Figure 18: Special Seven Filling Terry Design with Two Pick Groups and Full Beat Up

The loop formation system with full electronic control lets you alter the height of the loop by accompanying the electronic weft ratio variator device on jacquard looms to program different weft ratios like 3-pick terry, 4-pick terry and so. By this method, different heights of loops can be achieved in the same shed.

CHAPTER VI

WET PROCESSING AND FINISHING OF TOWELS

Terry towels are made up of terry fabrics, which are manufactured in power looms with a pile yarn in the warp, in addition to the ground yarn which goes in to the base fabric along with the weft yarn inserted transverse during weaving. The piling (loops) of the yarn imparts a furry appearance to the fabric surface and creates a lot of additional surface area. Being woven out of 100% cotton yarn, these fabrics acquire significant hygroscopic property. This is the prime motivation behind using terry fabrics in home textiles in general, towels and bathrobes in particular. Terry towels come in different sizes and shapes. Some of them are intended for use in a particular setting e.g. beach towel or for a particular purpose e.g. kitchen towel, baby hood towel, or in sizes convenient for a special use, like hand towels, face towels, bath sheets etc. Among other important products in the home textiles segment is the "mat". These are typical adsorbent mats meant for the dry European washroom. These are also made with terry fabric, but in higher GSMs by achieving higher pile density with yarns of lower counts. The typical products are „bath mat" and „tub mat", which are under a continuous demand.

Product Design

Almost always, the towels are manufactured according to certain design specification given by the customer or on designs developed in house and approved by the customer. The specifications relating to the product, which include Product requirements, are received from marketing through new enquiry.

Weaving: As per requirements are given by PPC, Terry Fabric of particular design is woven and sent to Frey Folding.

Grey Folding: As stated earlier, the fabric produced by the weaving department, on each roll is sent to the next department known as the Grey Folding Department. Before sending into the Grey folding department, a computer data entry of the details of the fabric on roll is made for the weaving and Grey-folding department.

The details entered into the computer are similar to the data entered in the Loom Sort Particular and in addition to that new data are also entered:

- Order No
- Sort No/Design No
- Length of the Fabric Roll

- Weight of the Fabric Roll both net gross weight
- Total No of pieces produced per size.

Grey Storage Room/Grey Folding Room

The objective of the Grey-folding is:

- To inspect the fabric received from weaving, whether Grey or Yarn dyed.
- To generate quality report of the fabrics inspected and intimate to the responsible department in case of problems or defects in the fabrics.
- To book the quantities of fabric that is produced per design per colour for each loom
- To update record order wise Vis a vis production and ordered qty
- To prepare the fabric for delivery to the processing department for the various processes as instructed by the PPC department.
- To store the gray goods in roll form, until programmed for issue to the processing.
- To return the empty cloth rolls to the weaving department.
- To mend all the mendable defects prior to rolling and storage

The fabric rolls are inspected on the inspection machine, which are provided with glass tops and light source to look at the various possible defects.

The fabric rolls are inspected and checked on the inspection machine for the following parameters:

- Sort No.
- Order No.
- Loom No.
- Size of the towel as against the reference towel
- Finished size of the towel in Grey form
- Pile height measurement by the pile ratio method
- Weight of the towel
- Visible defects/weaving defects that can lead to rejection in the final stage
- Total pieces produced in that roll

The Grey rolls sent to processing undergo following processes. Based on the processing needs as instructed by the Production Planning and Control, the Grey Folding Department arranges to move the fabric into the processing region along with a job card. The Grey issue program is given by the PPC to the Grey-folding department after update, as and when the fabric rolls are inspected.

Terry Fabric Dyeing: In the processing department Dyeing and Finishing of the Terry Fabric is done. These are various routes followed in a typical towel process house:

These routes are very important and are based on the Buyers requirement of the Finished

Terry Towel Products and generally a successful proven route is followed by the Process House. These are cost specific and any change in the routes is adding to the cost of Processing. All technical staff, therefore, is to be given very good understanding of these routes to avoid any damage to the fabric.

Washing of towels: This is usually done in case of yarn dyed towels, where the fabric is just washed before sending it to the Finish Folding department.

Dyeing: Grey fabric is dyed into solid colour towels.

A typical recipe and duration of the process is could be:

Route No.	Route Description	Machines-involved
1	Piece Dyed	Dyg M/c-Hydro extractor-Rope Opener-Tumble Dryer-Stenter
2	Dyed Velour	Shearing-Dyg M/c-Hydro extractor-Rope Opener-Tumble Dryer-
3	Yarn Dyed Velour	Shearing
4	Yarn Dyed Velour-Wash	Shearing-Dyg M/c(Wash)-Hydro extractor-Rope Opener-
5	Yarn Dyed-Washed	DygM/c(Wash)-Hydroextractor-Rope Opener-Tumble Dryer-Stenter
6	Greige-Velour	Shearing

1	WHITE	
S. NO.	NAME OF CHEMICALS	CONSUMPTION %
1	Wetting Agent	1.000
2	Desizing Agent	0.500
3	Sequesting Agent	0.200
4	Lubricating Agent	0.500
5	Caustic	2.000
6	Peroxide Stabilizer	1.000
7	Hydrogen Peroxide	8.000
8	Peroxide Killer	1.000
9	Acetic Acid	1.000
10	Softener	
	Cationic	2.000
	Silicon	2.000
11	Softener(Special For Optic White)	
12	Optical Brightner	1.200

2	LIGHT SHADE	
SL NO.	NAME OF CHEMICALS	CONSUMPTION %
1	Wetting Agent	1.200
2	Desizing Agent	0.500
3	Sequesting Agent	0.400
4	Lubricating Agent	1.000
5	Caustic	2.000
6	Peroxide Stabilizer	0.800
7	Hydrogen Peroxide	2.000
8	Peroxide Killer	1.000
9	Acetic Acid	2.000
10	Dye Bath Controller	0.500
11	Salt	20.000
12	Dyes	
13	Soda Ash	10.000
14	Detergent	0.500
15	Softener	
	Cationic	2.000
	Silicon	2.000
16	Softener(Special For Optic White)	
17	Optical Brightner	1.200

3	DARK	
SLNO.	NAME OF CHEMICALS	CONSUMPTION %
1	Wetting Agent	1.200
2	Desizing Agent	0.500
3	Sequesting Agent	0.400
4	Lubricating Agent	1.000
5	Caustic	7.000
6	Peroxide Stabilizer	0.800
7	Hydrogen Peroxide	2.000
8	Peroxide Killer	1.000
9	Acetic Acid	2.000
10	Dye Bath Controller	1.000
11	Salt	80.000
12	Dyes	
13	Soda Ash	15.000
14	Detergent	2.000
15	Softener	
	Cationic	2.000
	Silicon	2.000
16	Softener(Special For Optic White)	
17	Optical Brightner	1.200

Note: The time temperature and instrumentation controls are decided as per Processors experience and as recommended by the machine manufactures. The Software control is the heart of Processing Department and under sole control of the Head. Most modern Machines are also having a logical control system of all dyeing machines in the process house (e.g. Orgatex for Thies. Fong, Schlavos and others manufactures have their own systems.). They also give instant recipe cost, and the Dyes and Chemical Stock Position. They help greatly in Inventory control of these chemicals and indicate possible Reordering level thereby reducing the procurement cost

The fabric Dyeing and Finishing is carried out in Thies Soft Flow Dyeing machines in Rope form and then the dyed fabric is Hydro extracted, opened in Full Width and then Dried.

Yarn Dyeing: In case of Yarn Dyed towels, the Yarn is received in Yarn dyeing as Grey and is dyed in required shade

Eco Block–Yarn dyeing-Based on the yarn requirement received from the weaving preparatory department, which is calculated on the basis of the pile length and the ground length, the total yarn requirement based on the sort number, design number and order quantity are given to the yarn dyeing department.

The Yarn Dyeing department carries out yarn dyeing for the sampling as well as for bulk production. Typical Yarn Dyeing machines are Stock Dyeing machines where the carriers are suitable for Yarn Dyeing. The required Hydro extractor is also suitably designed to dry the dyed Yarn packages(Cheese).

The dryer is also a suitably designed Steam Package Dryer OR a Radio Frequency Dryer. The dried packages are then rewound on Paper cone to be sent to Preparatory department.

Towel Finishing: Mostly Terry Towel Fabrics are applied with various chemical and Mechanical Finishing Chemicals and then Hydro Extracted and Dried. Textile markets are very strongly governed by the finish of the material. Day by day demand of specialty finishes is coming in market and the processor has to cater these in order to make survive his unit.

These finishes are as per Buyers requirements and their dosages are dependent on the requirement and are decided as per the experience of the Processor.

1) Soft/Crispy/Silky Soft Finishes: There are various demands of the finishes from market and the Softness, and Handle are very typical of a particular order and the required dosages are strictly followed for that order. Generally Cationic and Silicon finishes are applied post dyeing in the Dyeing Machine itself. The drying is kept under control.

2) Antimicrobial: For odourless effect after repeated use of towels certain specialty chemicals are applied on fabric

3) Stain Release: When the required chemicals are applied the towels acquire property to release the stain after washing easily.

4) Aroma: In order to impart pleasing fragrance to the towels some good aromatic chemicals are imparted.

5) Lasting Colour: Along with the use of select dyes from the manufacturer certain finishes are applied to make towel retain its colour brightness till many washes.

6) Ever bright: A select brand of Optical Brighteners along with use of prescribed chemicals makes white towel retains its whiteness.

7) Lint Free: There is very tight control of the Government in US and EU on the household waste coming out of washing machine. To limit the lint coming out of Towel washing, a suitable choice of Yarn is carefully made and some specialty chemicals are applied. A study is made after 10,20,50,80 washes and a lint report after washing is made. This report increases the salability of Towels there.

8) Quick Dry: Along with the use of change in fibre composition and yarn morphology, certain chemicals when applied, towels loose moisture easily and dried quickly. This saves a lot of Household Electric energy thereby reducing the expenses.

CHAPTER VII

PHYSICAL PROPERTIES OF A TERRY TOWEL

1. Absorbency

High absorbency can be achieved in a towel by increasing the surface area with pile yarns and using cotton yarns with twists lower than the ground warps.

2. Heat Insulation

Pile yarns make the fabric thicker and give the fabric a high level of heat insulation. Moreover cotton fibers which are used in towels are naturally convoluted and bulked. This serves to trap air within the fabric structure. The air contained between fibers and within them provides thermal insulation. These convolutions plus the tapered fiber ends also hold the fabric away from the skin, adding to the amount of air trapped and contribution to heat insulation. According to the results of an experiment which was carried out by Morooka, dry heat loss of toweling fabrics was found to be lower than that of common cotton fabrics on the market. However dry heat loss was found to be higher than is expected from the thickness and apparent density.

3. Crease Resistance

Pile yarns give the fabric a third dimension which makes the fabric nearly increasable.

4. Dullness

The pile loops form a very rough textured surface, thus giving the fabric a dull appearance. This situation is true for only un-sheared toweling. Velour toweling has an appearance even brighter than that of a traditional fabric. The cut pile forms a very smooth surface and reflects light evenly. The pile direction on both velour and uncut terry fabric also has an effect on the color appearance. This is related to the reflection angle which changes with the pile direction. This effect is more obvious in velour terry towels. When the pile direction is laid downwards, the fabric offers a smoother surface for light and so appears more lustrous. If the pile is erect, the color is richer because more of the fabric (and color) is visible while looking into the depth of pile loops.

CHAPTER VIII

ADVANCES IN MACHINES FOR PRODUCTION OF TERRY TOWEL PRODUCTS

1. Stitching Automation-Monti Mac, Italy

Monti Mac was established in 1931, and provides automated systems for fabric preparation, sewing, cutting and confection processes for fabrics. For stitching of terry to welfabric, Monti Mac offers automatic longitudinal and bilateral hemming and automatic cross cutting and cross hemming machines.

2. Terry Towel Pretreatment & Dyeing-Toweltec by Fong's Hong Kong

One of the world"s leading manufacturers in the field of processing, Fong"s Industries" primary business is the design, manufacture, and sale of equipment for yarn and piece dyeing, dryers, and finishing.

3. Toweltec–2T High Temp High Pressure Soft Flow Dyeing Machine

Fong"s TOWELTEC is a high-temperature, high pressure, piece dyeing machine. Featuring an innovative nozzle design and lint collecting system(that automatically filters,accumulates, and discharges lint); the TOWELTEC machine ensures highly efficient and eco-friendly pretreatment& dyeing. The unique nozzle and internal plaiter is designed to keep the fabricina relaxed and fluffy state, effectively reducing loop disorders.

Designed for high capacity production without sacrificing efficiency, this machine is intended to ensure high repeatability and levelness. The TOWELTEC range of machines is available in single chamber to multiple chamber (up to 6 chambers; each chamber with a capacity of 400kg) variants.

"Mainly used for zero twist towels and different treatment capabilities, the FONG"s machine has considerably helped us to balance the gap that existed in the system we used previously to achieve higher production targets. We have been running the machine at full capacity for almost a year now and the machine hasn"t troubled us for a single moment", says Mr.Abhishek Shrivastava, Plant Head of Oracle Exports, Vapi.

Reference: Oracle Exports(Vapi), Noman Terry Towel (Bangladesh), Loftex Industries, Sun Vim Group(China)...

4. Colour Kitchen–Eliar, Turkey

Since its inception in 1984 in Turkey, Eliar Elektronik San A.S. has focused on automation systems for fabric and yarn dyeing machines. Today, Eliar offers components and complete system solutions for all types and brands of dyeing and processing machines.

In addition to providing finishing solutions, Eliar also offers multiline and monocline chemical & colour dispensing systems, powder dyed is solving systems, salt and soda dissolving/distribution systems as well as dry powder distribution in 1:1 sludge form and laboratory dispensers (for labtrials).

5. Rope Opener–Corino, Italy

Corino, Italy, has vast experience in rope opening and slitting machines that ensure smooth and hassle-free running of fabric. For terry towel applications, the Corino rope opener can be installed in front of the tumble drier.

6. Montexstenter–A. Monforts, Germany/Monfongs, Hong Kong

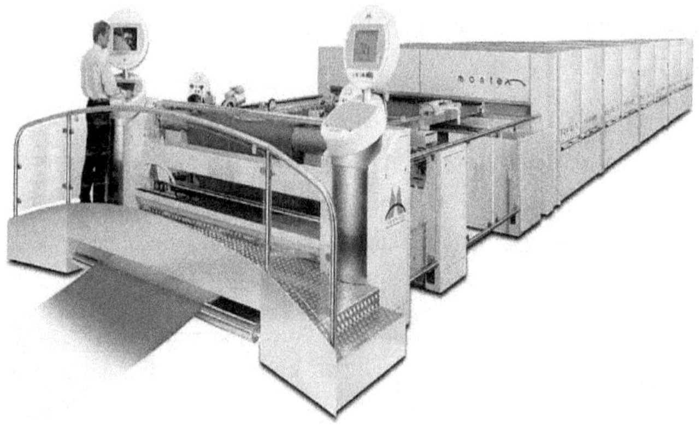

Founded in 1884, A.Monforts is the market leader for a complete set of processing machinery including dyeing, finishing ranges, continuous dyeing ranges, relax dryers, shrinking ranges, and the like. Monforts" Montex St enter features technologies that ensure economical and eco-friendly operation. In general, to ensure final width is achieved, after the fabric is passed through the tumble drier, the fabric is again passed through the stenter. In addition to setting the width of the fabric, this process also ensures that the desired residual moisture is achieved as well. These stenters are designed in Germany and produced in Germany or China.

7. Weft Straightener/Pattern Control System–Mahlo, Germany

MahloGmb Hand Co., Germany, is the world's leading manufacturer of weft straightener, process control and monitoring automation systems. Mahlo"s process control equipment allows manufacturers to achieve high productivity, cost efficiency, resource allocation optimization, consistent and repeatable quality with short lead-times.

Mahlo"s weft straighteners are high-quality, automatic bow & skew detection even in complex structures and correction devices that ensure the precise production of high quality fabric with reduced costs. These state-of-the-art weft straighteners are highly accurate: they can straighten even min or degrees of distortion.

In the case of terry towel fabric, it used to be difficult to detect weft yarn due to the pile of the fabric. However, Mahlo-"s unique PCS-12(Pattern Control System) allows for online contactless optical scanning, monitoring, and control of pattern distortion(both bow and skew) with a system of cameras.

8. Continuous Bleaching/Pad Steam Dyeing/Washing Range–Goller, Hong Kong

For more than a century, Goller has been at the fore front of the development, manufacturing, and supply of wet finishing systems around the world. Goller offers a wide variety of efficient and well-designed machinery for woven/terry fabrics and knits from desizing to washing. Bulk textile manufacturers prefer to use continuous bleaching and pad-steam reactive dyeing processes, both of which Goller has vast experience in.

For terry towel processing Goller offers the following machinery:

Goller Complexa: designed to have a modular construction, the Goller Complexa bleaching range allows for pre-treatment of terry fabrics for scouring and bleaching, and also two-step bleaching process.

Goller Colora: Goller Colora is a pad steam dyeing range that can be configured for dyeing of reactive colours or development of vat colours.

9. Flatbed/Rotary/Digital Printing–Zimmer, Austria

Based in Austria, Zimmer Maschinenbau GmbH is a world-class manufacturer of finishing machinery for textiles, carpets, non-wovens, paper, and other roll-to-roll substrates. In the terrytowel sector, Zimmer offers the following digital/flatbed printing products:

Zimmer"s Rotascreen rotary screen printing machine with the Zimmer Magnet System Plus can be sometimes used for printing of terry fabric.

Digital Printing Machines from Zimmer   Flatbed Printing Machine from Zimmer

Zimmer"s Magnoprint flatbed printing machine is a proven product that features a unique magnetic system and roll rod technology in the longitudinal direction that enables single or multi-colour printing. One of the most versatile flat screen printing machines available today, Magnoprint allows for step less adjustment of the length, speed, and number of the squeegee stroke. Magnoprint also features the original Magnet System Plus, for magnetic roll rod and magnetic squeegee printing in the direction of the warp. For towel printing applications, the magnet bars used are stronger. Zimmer"s Colaris digital printing machine is a high capacity, industrial inkjet printing system for textile applications that guarantees low printing costs. An operator-friendly machine, Colaris is designed to enable a single operator to manage several machines. Unlike traditional printing machines, a print head failure does not lead to a halt in production; production continues, albeit at a reduced speed. The Colaris digital printing machine is even suitable for heavy sheared terry towels as well with its patented in-line pretreatment system–MAGNOROLL which allows wet on wetprinting. The only digital printing system for pile heavy terry products and other pile fabrics, Zimmer"s ChromoJET features powerful jets of premixed spotcolours & simulated or semi processed colours that inject the dye deep in to the fabric so that the desired level of penetration is achieved. ChromoJET also works with a variety of dyestuff groups–reactive, acid, disperse, pigment, etc, thus ensuring maximum efficiency coupled with low cost and high fastness. Featuring a modular concept, Zimmer"s ChromoJET allows users to extend the number of colours and jets from 8 colours/64 jets upto16 colours/256 jets–thus increasing flexibility and productivity.

Reference: Flatbed-Bhavik Terryfab(Jaipur), Digital-Loftex Industries (China), Loop steamer–SalvadèS.r.l., Italy.

SalvadèS.r.l. offers after-print steaming systems, and dryers. Salvadè„s revolutionary Super Wetsystem (patented) is used in the Loop Steamer"s chambers. The Loop Steamer features continuous fabric transport throughout the machine to prevent rubbing. The steamer chambers are maintained at a constant temperature throughout. The machine can also be heated quickly from 0-180°C.

10. Metal Detector–CEIA, Italy

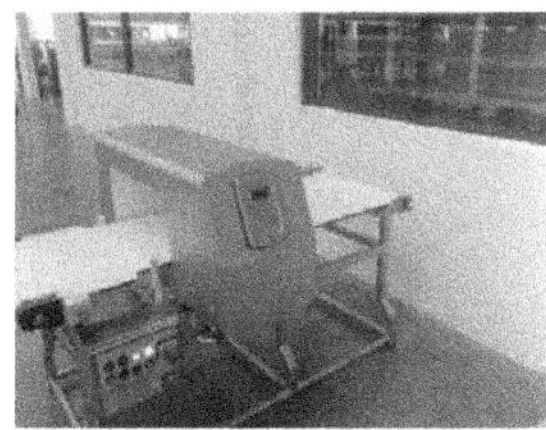

CEIA electronic metal detectors are universally recognized by leading world manufacturers of industrial machinery for their high quality and reliability. CEIA offers industrial metal detectors which are the ideal means of protection for production lines against accidental damage caused by pieces of metal which can enter the manufacturing process along with the material. CEIA offers both bar-type metal detectors(for detection of metal particles in open width fabric) as well as conveyor-type metal detectors(for fabric in packed form).

Reference: Welspun(Vapi), Trident(Ludhiana), SEL (Ludhiana) Effluent treatment plants–A.T.E. Envirotech, India.

A.T.E. Enviro tech offers a comprehensive range of technologies to address the individual treatment and recycling needs of textile manufacturers. These technologies include zero liquid discharge, Dissolved Air Flotation (DAF) devices, an aerobic systems for reduced overall power consumption, bio-enzyme treatment for eco-friendly and efficient degradation of complex organics, colour and refractory Chemical Oxygen Demand(COD) and reducing sludge generation, etc. A few salient features of A.T.E. Envirotech"s ETPs areas follows:

- Systematically designed unit operations and processes.
- Robust designs that conform to international standards.
- Engineered for low operating cost and smaller footprint.
- Optimized instrumentations and controls for fail-safe operations.
- Simplified operations backed by comprehensive monitoring and maintenance services.

CHAPTER IX

QUALITY TESTING OF TERRY TOWEL AND ITS ASSURANCE

1. Introduction

The testing of quality of textile products is an expensive business. A laboratory has to be setup and furnished with arrange of test equipment not just when results are required.

More overall these costs are nonproductive and therefore add to the final cost of the product. Therefore it is important that testing is not undertaken without adding some benefit to the final product. There are a number of points in the production cycle where quality testing may be carried out to improve the product or to prevent sub-standard merchandise progressing further in the cycle.

Reasons for Textile Testing:

1. Checking Raw Materials
2. Monitoring Production
3. Assessing the Final Product
4. Investigation of Faulty Material
5. Product Development and Research

1. *Checking Raw Materials*

The production cycle as far as quality testing is concerned starts with the delivery of raw material. If the material is incorrect or sub-standard then it is impossible to produce the required quality of final product. The textile industry consists of a number of separate processes such as natural fibre production, man-made fibre extrusion, wool scouring, yarn spinning, weaving, dyeing and finishing, knitting, garment manufacture and production of household and technical products. These processes are very often carried out in separate establishments; therefore what is considered to be a raw material depends on the stage in processing at which the testing takes place. It can be either the raw fibre for a spinner, the yarn for a weaver or the finished fabric for a garment maker. The incoming material is checked for the required properties so that unsuitable material can be rejected or appropriate adjustments made to the production conditions. The standards that the raw material has to meet must be set at a realistic level. If the standards are set too high then material will be rejected that is good enough for the end use, and if they are set too low then large amounts of inferior material will go forward into production.

2. *Monitoring Production*

Production monitoring, which involves testing samples taken from the production line, is known as quality control. Its aim is to maintain, within known tolerances, certain specified properties of the product at the level at which they have been set. A quality product for these purposes is defined as one whose properties meets or exceeds the set specifications. Besides the need to carryout the tests correctly, successful monitoring of production also requires the careful design of appropriate sampling procedures and the use of statistical analysis to make sense of the results.

3. *Assessing the Final Product*

In this process the bulk production is examined before delivery to the customer to see if it meets the specifications. By its nature this takes place after the material has been produced. It is therefore too late to alter the production conditions. In some cases selected samples are tested and in other cases all the material is checked and steps taken to rectify faults. For instance some qualities of fabric are inspected for faulty places which are then mended by skilled operatives; this is a normal part of the process and the material would be dispatched as first quality.

4. *Investigation of Faulty Material*

If faulty material is discovered either at final inspection or through a customer complaint it is important that the cause is isolated. This enables steps to be taken to eliminate faulty production in future and so provide a better quality product.

Investigations of faults can also involve the determination of which party is responsible for faulty material in the case of a dispute between a supplier and a user, especially where processes such as finishing have been undertaken by outside companies. Work of this nature is often contracted out to independent laboratories that are then able to give an unbiased opinion.

5. *Product Development and Research*

In the textile industry technology is changing all the time, bringing modified materials or different methods of production. Before any modified product reaches the marketplace it is necessary to test the material to check that the properties have been improved or have not been degraded by faster production methods. In this way an improved product or a lower-cost product with the same properties can be provided for the customer. A large organization will often have a separate department to carry out research and development; otherwise it is part of the normal duties of the testing department.

2. Basic Parameters of Quality Terry Towel

- Weight & GSM: Weight and GSM should be same as required by customer. Every manufacturer has some template or software(ERPs) where towel manufacturers calculate everything like spile"s height, density of picks and ends to meet requirement. This database or any software has been developed through some basic calculation.

- Softness/Hand feel: It depends on properties of the yarn used in pile, finishing chemicals and too some extent on pile orientation.

- Pile Orientation: Totally depends on process line.

- Lint: Lint is basically protruding fibers present in a finished towel. It is measured by weight of accumulated fiber collected from washing machine and tumble drying machine during testing.

- Absorbency: Terry towel should be highly water absorbent.

- Dimensional Stability: How a towel is behaving after washing is fall under dimensional stability properties. Dimensional stability is measured by the residual shrinkage% in a finished towel.

- Other Parameters are strength, colorfastness etc.

3. Methods

The conditioning and testing of textiles instances where such conditioning is specified in a test method. Because prior exposure of textiles to high or low humidity(65+/-2%.) may affect the equilibrium moisture pick-up, a procedure also is given for preconditioning the material when specified. The equipment to be used in the conditioning and testing of textiles shall include conditioning room or chamber, preconditioning cabinet, room, or suitable container, balance, and multiple shelf conditioning rack. The conditioning room or chamber shall consist of equipment for maintaining the standard atmosphere for testing textiles throughout the room or chamber within the tolerances given and including facilities for circulating air overall surfaces of the exposed sample or specimen and equipment for recording the temperature and relative humidity of the air in the conditioning room or chamber.

Samples or specimens requiring preconditioning shall be brought to relatively low moisture content in a specified atmosphere. Samples or specimens requiring conditioning shall be brought to moisture equilibrium for testing in the standard atmosphere for testing textiles, or when required.

4. Water Absorption Test

This is most important test carried out for terry towel. Main use of terry towel is to absorb the water after bathing. So to check the water absorbency is necessary. This test method determines the ability of a terry fabric to rapidly absorb and retain liquid water from surfaces such as human skin. The sample to be tested is mounted on an embroidery ring with just sufficient tension for removing the wrinkles in the fabric. It is placed below burette with the tip of the burette not more than 2.5cm above the fabric. The fabric is placed horizontally. The burette is filled with distilled water upto the zeromark. A drop of water is then allowed to fall on the fabric and the same time a stopwatch is started. The drop is viewed at a low angle and as soon as the light from the drop vanishes, the stop watch is stopped and the time in seconds is noted. The test is repeated in other portions of the same towel ten times and average is calculated. The time so calculated is recorded as the absorbency of fabric.

5. Dimensional Stability Test

To determine dimensional change of the towel comparing the distances between length and width direction benchmarks before and after when subjected to an appropriate combination of specified washing & drying procedure. ECE(European Colorfastness Establishment) detergent (without BOA), sodium perborate tetra hydrate used. Lay the towel to relax for 4 hours on a bench in ambient conditions so that it is smooth & tension free. Select the correct programmed for the wash required depending on the type of machine being used, set temperature, water levels. Dissolve the powder with a small quantity of warm water in a beaker and start the machine. Tumble dryer used to dry the towel.

6. Fastness Properties of Terry Towel

Fastness there is resistance of dyeing and prints to external influences, is having importance for the practical use of any dyestuff. Fastness property tests are consequently very extensive and widely standardized. Technical committee 38, sub-committee of the International Standards Organisation(ISO) has to date defined 45 different methods. The purpose for which a dyed material is to be used must always be born in mind when dyestuffs are selected. The evaluation of its results, effected with the grey scale used to determine changes of shade and staining.

7. Fastness to Washing

A 10x4cms watch of the dyed fabric is taken and is sandwiched between two adjacent (grey cotton) fabrics and stitched. The sample and the adjacent fabric were washed together. Five different types of washing are specified as different washing methods.

The solution should be preheated to the required temperature of washing. The liquor ratio should be 1:50. After soaping treatment, the specimen is removed, rinse twice in cold water and then in cold running tap water. Squeezed and dried in air at a temperature not exceeding 60°C. Place the tested sample next to a sample of the same material which has not been tested, and compare the change with the help of grayscale under good lighting conditions and give the grades. If the dyeing being tested shows equal or less change than the satisfactory sample, then it is as good as the satisfactory sample. Place the adjacent fabric next to samples of the same material which have not been tested and compare them. Equal or less staining shows equal or better fastness.

8. Fastness to Light

The purpose of Colorfastness to lightest is to determine how much the color will fade when exposed to a known light source. It is an offline quality assurance system. Generally in India to Wels kept outside of the home for drying purpose. Inday; sunlight fall on the towel surface. So it needs to know how much protection ability have a fabric to sunlight. It is determined by an experiment called colorfastness to light.

Principle of Color Fastness to Light

This test measures the resistance to fading of dyed textile when exposed today light. The test sample is exposed to light for a certain time which is about 24 hours to 72 hours or by customer/buyer demand and compare the change with original unexposed sample the changes are assessed by Blue Scales.

Light Fastness Grades

Grade	Degree of Fading	Light Fastness Type
8	No fading	Outstanding
7	Very slight fading	Excellent
6	Slight fading	Verygood
5	Moderate fading	Good
4	Appreciable fading	Moderate
3	Significant fading	Fair
2	Extensive fading	Poor
1	Very extensive fading	Verypoor

9. Fastness to Rubbing

A fastness is a place, such as a castle, which is considered safe because it is difficult to reach or easy to defend against attack. This test is designed to determine the degree of color which may be transferred from the surface of a colored fabric to a specify test cloth for rubbing (which could be dry and Wet).

There are two test methods for rubbing fastness

1. ISO-105-X12
2. AATCC-08

In ISO-105-X12 the wet pickup of the rubbing cloth is 100%.While in AATCC-08 the wet Pickup of the rubbing cloth is 65%. We check rubbing by Dry and Wet methods. In wet rubbing we wet the rubbing cloth according to test method and give rating by comparing the Staining with the grayscale. Similarly for dry rubbing we check the rubbing with dry rubbing cloth and compare the staining with grayscale for ratings. Color Fastness to rubbing is a main test which is always required for every colored fabric either it is Printed or dyed. If the color fastness to rubbing is good then its other properties like Washing fastness and durability etc improves automatically because the rubbing is a method to check the fixation of the color on the fabric. So if the fixation is good its washing properties will be good.

Rubbing Fastness depends on

- Nature of the Color
- Depth of the Shade

Construction of the Fabric Nature of the color each color either it is pigment, Reactive, Disperse or direct has its own fastness properties to rubbing. There are some colors like black, Red, Burgundy, Navy blue which have poor Colorfastness properties because of their chemical structure. Like Black color is a carbon base color and the particle size of carbon is large than the other colors that's why its rubbing properties are poor. Similarly red and blue are in the same case. So to improve the colorfastness we add more binder to improve the fastness properties of these colors. It doesn't mean that we cannot achieve the best results with these colors. The required results can achieve but production cost will be increase. On the other hand the construction of the fabric also affects the fastness properties.

10. Always Check

- Quality construction
- Color
- Depth of the Color
- End Use of the product

Results which we can achieve in Normal Conditions are

Dark Shade	Medium Shades	Light Shades
Dry 3-4	4	4-5
Wet 2-2.5	3	3.5-4

CHAPTER X

DEFECTS THEIR CAUSES AND REMEDIAL MEASURES IN TERRY FABRIC

Introduction

The manufacturing of terry towels/fabrics demand for specialize knowledge to avoid the pitfall those are inherent and which causes many problems, right from the weaver to head of department as well as sales staff too. Fabric defect can be defined as an unacceptable textural difference, caused by many factors, that undetermined the quality of fabric. Fabric quality can be quantified as the number of defects per square meter. One of the most elusive and confusing problems faced by textile technologists is the analysis of the fabric defects. To analyze the cause of defect a textile technologist has to make a guess mate of "What could and where had happened" situation to put forth his hypothesis. It may or may not be true reflection of a situation. Properties of raw material, processing deviation in terms of man-machine settings and their interactions with the material, all can contribute to the problem of defect generation. According to a report, fabric faults or defects are responsible for nearly 85% of the defects encountered during the manufacture of garments. Manufacturers recover only 45% to 65% of their profits from seconds or off-quality goo Identification of fabric defects offers both a challenges as well as opportunity. Challenge comes in the form of pinpointing the exact stage at which the fabric defect got inducted as well as its contributing factors. It provides opportunity to enrich analytical ability and joy of finding the "solution", on one hand and to eliminate or at least restrict to a minimum the occurrence of these defects on the other hand. One usual trouble occurring especially when weaving reversible patterns in two tones, or white and color terry, where one color shows on the face, and the other color on the back side of the terry fabric, or alternately changing from face to rear, is the tendency of color to mix into the part, where only white terry should appear or where color should appear. The smallest spinning slub or knot would to drag back the pile yarn from the loops previously inserted during weaving, and either stops the loom because of the slackened end or make a slight tangle. On the way or another it either causes an annoyance to the weaver or disturbs the overlay of the terry so as to produce "seconds". No appreciable gain in production, possibly a small loss. Terry fabrics and terry looms/terry weaving machines are subject to most of the problems and the faults those are associated with the weaving of flat fabrics, but there are some additional defects, associated specifically with the three or more pick cycle of terry weaving. The defects which can be found in toweling are shown on the following table

Table 4: Quality Defects which are Seen on Terry Woven Fabrics

Weaving Defects	Weft Processing Defects	Sewing Defect	General Defects
Missing Pile	Uneven Dyeing	Low stitch No.	Crushed Pile
Missing Filling Pile	Off shade	Widths of Parts are out of Tolerance	Stain
Thin or Thick filling Pick	Print Defect	Missing or Faulty Stitch	Cut, hole, tear, or burst
Beginning or End part Missing	Design Defect	Un-reinforced Stitching on Ends	Any dimensions out of standard
Side Part is Missing	Noxious or Offensive Odors		
Stop Mark			
Reed Mark			
Wavy Selvedge			
Dense Weft			

Dense weft defect means that the higher density of the border part is used by mistake in the pile part. Out of tolerance parts' widths mean the total width and the folded-in width of the beginning, end and side parts are lower than required by the standard.

Problems Caused Due to Weft Breakages

If the loom does not stop "on pick", when a weft break occurs then because of the basic three pick terry structure, it may be necessary not simply to remove, and replace the pick but to remove and replace the whole group of picks so as not to produce faulty or abnormal loops. All picks must be pulled out to find the correct pick for a new start. The ground beam and pile beam are pulled back, but care must be taken when pulling back the pile beam not to pull back too far, or the terry loops left in cloth will be pulled out.

Formation of Cracks When Changing from Border to Pile

It is usually observed that a visible unsightly "crack" is formed across the full width of fabric when changing from border weave to pile weave but which does not occur when changing from pile weave to border. In order to prevent this defect, it is usual to insert two or sometimes more "cramming picks" into the border weave immediately before the change back on to terry weave takes place. The method of arranging the change is found to influence its severity. There are three different ways in which the fallback of reed can be made to operate when changing from border to pile, depending on the point at which the pattern changeover has been made from terry to border and similarly eventually back to terry weave.

- The reed falls back for two picks, and then beats up fully after the three picks, which causes the worst "crack".
- The reed beats up fully on the first pick then fallback on second and third picks, which gives a less pronounced "crack".

- The reed falls back on the second pick beats up fully on the second pick, and falls back on the third pick.

This gives the least pronounced "crack", and if this arrangement is employed in conjunction with the "cramming picks" the crack can be avoided. In first two arrangements, cramming will not cover the crack.

Formation of Random Warp Wise Cracks

Another type of "crack" that appears in terry fabric, is a gap in the warp direction in the pile, extending typically for 2 to 15 cm. In length and randomly distributed. This gives furrow like appearance in the terry fabric. Such appearances at places giving delusion of missing piles from these portions. This defect is seen more prominently in figured pile fabric than in plain pile fabrics.

Causes

In figured pile fabric, generally the furrows start at the point of crossover of a pile from face to back or from superficially correct practice of arranging the warp yarns in the reed in the way that gives the most uniform distribution of pile i.e. ground and pile alternately with one of each, in each reed dent. Such an arrangement causes the around warp to occupy the outer position in each dentin one half of the cloth. However, the natural tendency for the cloth to contract at the fell causes the ends to be drawn in wards and the around end being much tighter, tend to move past the pile warps causing them sometimes to change the position.

Remedies

1. To draw two ground ends to a dent, then two piles ends and soon.
2. To change the order at or near the centre of reed or shifting the reed side way relatively to the healds so as to give a one way bias across the whole width but both methods may give rise to the other problems. One of the major problems encountered is that when changing direction of pile for Jacquard figure effects, unsightly line appears in the warp wise direction. These furrows appear in random on the face and the back of the fabric, and are caused by the pile threads tending to migrate from one side to the other of adjacent ground ends.
3. In order to prevent unsightly warp wise cracks from appearing in counter changing patterns, it is necessary to separate the pile ends from the ground ends in the dent.
4. In order to prevent pile warp ends from migrating around each other, when adjoining

ends are forming loops on the same side of the cloth, i.e. lifting together, it is necessary to separate them from each other in the dents.

5. It is not necessary to separate the ground warp ends from each other, as they weave in opposing shed all the time, and do not migrate around each other.

Mixed Terry

This defect known by such names as mixed terry, mingling, dogs teeth etc, while ever present can easily be corrected and not be repetitive trouble. The loosely weighted pile end is likely to be adversely affected by anything, which it may come into contact.

Causes

1. The ground ends being heavily tensioned, and consequently tightly woven, must bear heavily against the sides of reed wires as the shrinkage from yarn spread to cloth width takes place during weaving when the pile end at the right hand side of its adjacent ground end in the dents of the reed, there will be problem. But that applies only to the right hand side of the fabric. The left hand end of the fabric will show mixed or ragged looking terry loops.

2. At this left hand side the ground ends are trapping the pile ends against the reed wires. As when the fast pick of the three picks forming the terry weave cycle, is beaten upto produce the loops, the imprisoned pile ends are nor perfectly looped. They mingled with the other ends, having loose and are pushed out of the position, often being interwoven wrongly, while hanging loosely the loop will go to the underside instead of the top a desired, or vice-versa, thus causing mixed terry or loose loops.

Remedies

1. To draw the pile ends at the right hand side of their adjacent ground ends in each dent in the right half of the towel. Somewhere near the centre, draw one pile–one ground–one pile (p-g-p) in one dent.

2. Reversing the last 9 inches while drawing a new set, and where necessary to increase this amount at the loom, on observing full effect.

3. In a full harness jacquard pattern fabric is to use the two and two method of drawing the pile and ground ends in the healds, Jacquard harness and reed. Then instead of a ground end another pile end is drawn through the next eye of Jacquard harness. Then draw first ground end through the eye in the healds haft and again another ground end in its respective eye of the ground healds. This across the full width of the yarn spread, the ends will lie "pile, pile, ground, ground" manner.

4. "Reversing the drawing" method only be used in good quality terry fabric with fairly fine reeds, and improves terry cover by drawing the ends in one dent. The pile ends are kept free from interference by the tight ground ends, the pile end being protected by the reed wires from the pressure of the ground.

Occasional Imperfect Loops: The ground ends, being drawn tight against the reed wires, may trap the pile ends occasionally not allowing them to be fully drawn forward into the loops. When this happens to a single end it doesnot influence the pile beam rotation, and delivery of yarn that is not consumed causes that end to be comes lack leading to loss of control by the healds or the formation of loops when there should not be one.

Ridgy Terry: The term "Ridgy terry" is used when whole rows of lops or substantial lengths of rows are formed with non-standard loop length, often alternate rows of different height. The fault is associated with poor pile tension control, and had been blamed on certain types of delivery mechanism. This fault will be referred to in relation to the terry loom.

Uneven Loops: Spiked ring temples are generally employed on terry looms. Correct setting is especially important because it is found that, if the temples are set too high or impede the movement of the cloth, result informing uneven loops near selvedges. An obvious possible reason for this is that with an inclined reed the effective fullback is a function of fell height. It is also known that the width wise contraction caused by crimping and hence stretching of the weft at beat up leads to slackness in this warp threads near the edges of the cloth and leads to the problem in warp control. Because of the high ground warp tension used in terry weaving, the weft crimp and cloth construction are greater that is normally the case, 10% contraction being a typical value.

Dark Streaks: This type of defect found in the case study of analysis of fabric defects carried out by the ATIRA. The fabric exhibit dark shade warp–way lines of varying width on face side having cut piles and were continuous in the given piece of fabric.

Causes

1. As the streaks were parallel and continuous, these would have occurred due to difference in characteristics of yarns used as pile warp. A group of caurse yarns appears lighter and group of finer yarns appears darker in the shade than the body of the fabric after dyeing.
2. One of the most prominent causes is the uneven dents pacing or wrong drawing-in of ends through the reed. Such streaks are relatively fine.
3. Even a variation nominal count of about 10% is sufficient to generate warp streaks if such yarns get grouped at warping and sizing.

4. Differences inlusture, reflectance or differential dye pickup of yarns due to difference in raw materials or blend composition can cause warp streaks.

Remedies

1. Proper drawing-in of ends from the reed and dent spacing should be done properly.
2. If possible use group of yarns of same characteristics.

Curled and Folded Surfaces

This defect is characterized by the appearance of curls and creases folds in the selvedges of the fabric after wet processing. Dyeing and printing are uneven in the vicinity of such creases. A selvedge, which gets curled and folded during wet processing, is often slack and wavy and exhibits a corrugated appearance at the grey stage. During wet processing such a selvedge gets further curled and folds itself at places, leading to crease formation after the fabric passes through the squeezing nip

Causes

1. Slackness of the selvedge is caused by incorrect choice of selvedge yarn count, twist, weave, drawing and denting order.
2. Selvedge folds are also sometimes caused by improper piece to piece stitching.

Remedies

Proper selection of yarn characteristics, weave, drawing and denting order reduce the chances of folding the selvedge.

Note: Formulae for Reproduction Calculations for Terry Towel

1. Production of loom (towel/per day/mc) =Rpm×60×24×efficiency/picks per towel
2. Finish weight= (Gsm × size of towel in cm)/10000
3. Lbs/doz= (finish weight × 12)/ 453.6
4. Piece weight = Gsm/size of towel in cm × (wt loss+100/100)
5. Picks in fancy = fancy size in cm × picks in fancy per cm
6. Total pick/ towel =length of towel+ plain border/cam --fancy size cm – fancy size cm ×picks/cm + picks in fancy × number of fancy borders
7. Width of grey towel in inches = pile ends per towel + ends in ribbon / half of reed
8. Length of grey towel in inches = length of towel in cm + plain border /2.54
9. Weight/Gm2= weight of towel in grams/ width of towel in cm/length of towel in cm × 10000
10. Loops in square inch = (picks per inch/3) + (half of reed +2)

11. Picks in fancy border = picks in fancy × no of fancy border

12. Pile ends/towel = size of towel in inches × 1.17 × half of reed

13. Ground ends/towel = pile ends + ends in selvedge

14. Wt of pile (gms) = Grey wt of towel in gm – ground wt. in gms – F.B wt in gms –weft wt.

15. Ground wt (gms) = ground ends/towel×length of towel in inch×1.14/36/843/ground count/2.2046/1000

16. Weft wt (gms) = total picks in towel–picks in fancy × (width of grey towel in inches+ 1)/36/840/weft count/2.2046/1000

17. F.B wt (gms) = picks in fancy × width of grey towel in inches/36/840/fancy border count/2.2046×1000

18. Reed space in inches = running towels/ machine × width of grey towel in inches

19. Reed utilization % = reed space in inches / max. Reed utilization per machine

20. Pile Ratio = pile weight in gms × 2.2046 × 840 × pile count × 36 / pile ends per towel/length of towel/1000

21. Pile height= pile ratio/2/ (picks per inch /3) × 25.4-1

22. Pile weight %= pile wt in gms/grey wt per towel in gms

23. Ground weight % = ground wt in gms / grey wt per towel in gm

24. Weft weight % = weft wt in gms / grey wt per towel in gms

25. F.B weight % = F.B weight/ grey wt per towel in gms

References

1. http://www.fibre2fashion.com/industry-article/textile-industry-articles/terry-towel.

2. some-developments-for-quality-improvement/terry-towel-some-developments-for-quality-improvement1.asp

3. V. Hobson, "Terry towels unraveled", Textile Horizons, Vol. 10, No. 4, Pp. 27-29, 1990.

4. M. Kienbaum, "Terry toweling-production techniques, construction, and patterning range", ITB, Vol. 1, No. 78, Pp.7-27, 1978.

5. N.M. Swani, P.K. Hari and R. Anandjiwala, "Performance properties of terry towels made from open end ring spun yarns", Indian Journal of Textile Research, Vol. 9, No. 3, Pp. 90-94, 1984.

6. M.D. Teli, Q. Munawar, S. Chaudhary and N. Saraf, "Finishing terry towel with softeners", International Dyer, Vol. 185, No. 4, Pp. 25-30, 2000.

7. U.K. Gangopadhyay, H.R. Vora, C.H. Sakharkar, R.A. Shaikh and V.A. Gawde, "Manufacture of terry towel in decentralized sector–a critical approach to enhance productivity and quality", Conference on Technological, Vol. 40, Pp. 138-144, 1999.

8. G.P.S. Kwatra, "Terry towel industry in India", Asian Textile Journal, Kwatra, Vol. 2, No. 3, Pp. 17-21, 1994.

9. http://www.ced-gujarat.org/pdf/Magazine/TERRY%20TOWEL.doc29

10. Annual Report2008–2009, Ministry of Textiles, GOI Source: http://www.domain-b.com/industry/Textiles/20091007_textile_ministry.html

11. http://textlnfo.files.wordpress.com/2011/10/terry-towel.pdf

12. http://www.ptj.com.pk/Web%202004/01-2004/seminar.html

13. http://www.zimmer-austria.com/cms/data/2/2013 ENGChromoJET_Terry_Towel_Printing_V1.pdf

14. http://www.ifc.net.au/edit/library_fin_dye_finishing/4.1.04%20KUSTERS.PDF

15. http://www.chasinggreen.org/article/better-bathroom-towels/

16. A. Al Faruque, S. Kader, S.U. Patwary, A.I. Mishuk, A.Y.M.A. Azim and S. Khatun, "Terry Towel in Bangladesh", European Scientific Journal, Vol. 11, No. 3, 2015.

17. N.D. Yilmaz, N.B. Powell and G. Durur, "The technology of terry towel production", Journal of Textile and Apparel, Technology and Management, Vol. 4, No. 4, Pp. 1-24, 2005.

18. R. Vogel, "Terry Fabrics with Exclusive Patterns", Sulzer Technical Review, Pp. 6-9, 1998.

19. M. Karahan, R. Eren and H.R. Alpay, "An Investigation into the parameters of terry fabrics regarding the production", Fibres & Textiles in Eastern Europe, Vol. 2, No. 50, Pp. 20-25, 2005.

20. I. Frontczak-Wasiak and M. Snycerski, "Use properties of terry woven fabrics", Fibres and Textiles in Eastern Europe, Vol. 12, No. 1, Pp. 40-44, 2004.

21. R. Subburaj, "Total Quality Management", Tata McGraw Hill Pub.Com. Ltd, New Delhi, 2009.

22. B. Purushothama, "Effective Implementation of Quality Management System", Wood head Publishing Ltd, New Delhi, 2010.

23. H. Lal, "Total Quality Management", New Age International Publishers Mumbai, 2009.

24. G. Vijaykumar and V.L. Sohani, "Design and Development of Computer based Training module for Total Quality Management in Textile Industry", The Bombay Textile Research Association Mumbai, 2004.

25. H.L. Needless, "Textile Fibers, Dyes, Finishes, and Processes", Noyes Publications, New Jersey, Pp. 34-40, 1986.

26. G.J. Cook, "Handbook of Textile Fibres: Part 1 Natural Fibres", Merrow Publishing Co.Ltd, London, Pp. 47-64, 1984.

27. W.E. Morton and M.A. Hearle, "Physical Properties of Textile Fibres", The Textile Institute, Manchester, 1993.

28. L. Hes, "Fundaments of Design of Fabrics and Garments with Demanded Thermo physiological Comfort", Textile Congress, Liberec, Pp. 94-95, 2001.

29. L. Hes, "Thermal comfort properties of textile fabrics in wet state", Proceedings of XI. International İzmir Textile and Apparel Symposium, 2007.

30. P. Yang and S. Kokot, "Thermal analysis of different cellulosic fabrics", Journal of Applied Polymer Science, Vol. 60, No. 8, Pp. 1137–1146,1996.

31. D.N. Saheb and J.P. Jog, "Natural fiber polymer composites: a review", Advances in Polymer Technology, Vol. 18, No. 4, Pp. 351–363,1999.

32. B.M. Prasad, M.M. Sain and D.N. Roy, "Properties of ball milled thermally treated hemp fibers in an inert atmosphere for potential composite reinforcement", Journal of Materials Science, Vol. 40, No.16, Pp. 4271–4278, 2005.

33. R.S. Chauhan, Vrunda Wala "Fabric Defects Causes and Analysis" Ahmadabad Textile Industry Research Association, Pp. 46-48, 2009.

34. Subhash J. Patil, "Manufacturing of Terry Towel", Pp. 591-596, ISBN: 81-85027-51-X.

35. http://textiletechnology,brarehost,com/spinning/barre.html

36. N.D. Yilmaz, N.B. Powell and G. Durur, "The technology of terry towel production", Journal of Textile and Apparel, Technology and Management, Vol. 4, No. 4, Pp. 1-24, 2005.

37. H.L. Needless, "Textile Fibers, Dyes, Finishes, and Processes", Noyes Publications, New Jersey, Pp. 34-40, 1986.

38. G.J. Cook, "Handbook of Textile Fibres: Part 1 Natural Fibres", Merrow Publishing Co. Ltd., London, Pp. 47-64, 1984.

39. W.E. Morton and M.A. Hearle, "Physical Properties of Textile Fibres", The Textile Institute, Manchester, 1993.

40. L. Hes, "Fundaments of Design of Fabrics and Garments with Demanded Thermo physiological Comfort", Textile Congress, Pp. 94-95, 2001.

41. L. Hes, "Thermal Comfort Properties of Textile Fabrics in Wet State", XIth International Izmir Textile and Apparel Symposium, Pp. 87-96, 2007.

42. P. Yang and S. Kokot, "Thermal analysis of different cellulosic fabrics", Journal of Applied Polymer Science, Vol. 60, No. 8, Pp. 1137–1146, 1996.

43. D.N. Saheb and J.P. Jog, "Natural fibre polymer composites: a review", Advances in Polymer Technology, Vol. 18, No. 4, Pp. 351–363, 1999.

44. B.M. Prasad, M.M. Sain, and D.N. Roy, "Properties of ball milled thermally treated hemp fibers in an inert atmosphere for potential composite reinforcement", Journal of Materials Science, Vol. 40, No. 16, Pp. 4271–4278, 2005.